Lecture Notes of the Institute for Computer Sciences, Social Informatics and Telecommunications Engineering   3

Maggie Cheng (Ed.)

Nano-Net

Third International ICST Conference, NanoNet 2008
Boston, MA, USA, September 14-16, 2008
Revised Selected Papers

 Springer

Volume Editor

Maggie Cheng
325 Computer Science Bldg.
500 West 15th Street
Rolla, MO 65409-0350, USA
E-mail: chengm@mst.edu

Library of Congress Control Number: Applied for

CR Subject Classification (1998): I.2.9, J.3, C.2, C.1.4, G.2.2

ISSN   1867-8211
ISBN-10   3-642-02426-2 Springer Berlin Heidelberg New York
ISBN-13   978-3-642-02426-9 Springer Berlin Heidelberg New York

springer.com

© ICST Institute for Computer Science, Social Informatics and Telecommunications Engineering 2009
Printed in Germany

Typesetting: Camera-ready by author, data conversion by Scientific Publishing Services, Chennai, India
Printed on acid-free paper   SPIN: 12649341   06/3180   5 4 3 2 1 0

Nano-Net 2008
Third International Conference on Nano-Networks

It gave me great pleasure to welcome everyone to this innovative conference and to the city of Boston on the third year since the vision of nano-scale networking was conceived. I was at the founding conference in EPFL, Switzerland and followed the second conference in Catania, Italy, with great interest.

We extended the Nano-Net conference this year to include a Workshop Session and a Tutorial Session; the highlight was, of course, the Technical Program. Alexandre Schmid did an outstanding job presiding over the review process used to select the papers that were presented at this conference. I am deeply grateful to the members of the TPC for their tremendous effort in evaluating (along with many anonymous reviewers) the submissions and for organizing the papers into sessions, all on a tight schedule. I extend my deepest thanks to the Organizing Committee, in particular, Alex Schmid and Karen Decker, as well as the active organizing members, namely, Sasitharan Balasubramaniam, Alexander Sergienko, Nikolaus Correll, Kaustav Banerjee, Radu Marculescu, and Tatsuya Suda.

The Nano-Net Organizing Committee selected three outstanding plenary speakers in Tatsuya Suda, Sylvain Martel, and Neil Gershenfeld, who spoke on a variety of truly fascinating topics regarding networking at the nano-scale.

The panel discussion on "Using Advanced Micro/Nano-electronic Technology to Establish Neuromorphic Systems" had very positive feedback. Panelists Garrett Rose, Vladimir Gorelik, Eugenio Culurciello, Shih-Chii Liu, and Matthew Hynd shared their perspectives on this growing technology related to the formation of the ultimate nano-scale network, a brain-like system.

Recent advances in nano-networks were presents in the form of a Poster Session. This opportunity encouraged informal discussion with presenters regarding their latest developments.

Karen Decker very ably served as Finance Chair and therefore was intimately involved in virtually all aspects of the planning and organization of the conference.

Neil Gershenfeld graciously handled local arrangements for the conference, including video recording of the presentations.

Yun Li organized an excellent tutorial. I would like to extend my warm thanks to Wei Lu for stepping up to share his ideas and expertise with the community in the form of this first Nano-Network Tutorial.

As you will see from the proceedings, Maggie Cheng has done an excellent job as Publications Chair.

Last, but far from least, I owe a special debt of gratitude to Sanjay Goel and Damira Pon of SUNY for an outstanding and highly professional job in organizing the Nano-Net Workshop and hosting the symposium website and responding to literally hundreds of requests for changes and updates to the website—all done in a timely and efficient manner.

I hope that you find the exchange of information that took place over those few fall days in Boston technically stimulating and professionally rewarding.

Stephen F. Bush

Organization

Steering Committee

Imrich Chlamtac (Chair)	Create-Net, Italy
Gian Mario Maggio	Create-Net, Italy

General Chair

Stephen F. Bush	General Electric, USA

Technical Program Chair

Alexandre Schmid	EPFL, Switzerland

Technical Program Co-Chairs

Alexander Sergienko	Boston University, USA
Kaustav Banerjee	UCSB, USA
Radu Marculescu	CMU, USA
Sumit Roy	U. Washington, USA
Nikolaus Correll	MIT, USA
Tatsuya Suda	UC Irvine, USA; DoCoMo, Japan
Sasitharan Balasubramaniam	TSSG, Ireland

Workshop Chair

Sanjay Goel	UAlbany, USA

Tutorial Chair

Yun Li	General Electric, USA

Panel Chair

Wei Wang	UAlbany, USA

Standards Chair

Dan Gamota	Motorola, USA

Publications Chair

Maggie Cheng   Missouri S&T, USA
email: chengm@mst.edu

Finance Chair & Conference Coordinator

Karen Decker   ICST

Webmaster

Damira Pon   UAlbany, USA

Applications for Nano-Networks

Sumit Roy   U. Washington

Modeling, Simulation, Standards and Architectural Aspects of Nano-Networks

Kaustav Banerjee   UCSB

Novel Information and Graph Theory Aspects of Nano-Networks

Radu Marculescu   CMU

Device Physics and Interconnects

Alexander Sergienko   Boston University

Nanorobotics

Nikolaus Correll   MIT

Nano-Biological Systems

Tatsuya Suda   UC Irvine, DoCoMo
Sasitharan Balasubramaniam   TSSG

General Technical Program Committee Members

Adriano Cavalcanti	CEO CAN
Alan Frieze	CMU
Alhussein Abouzeid	RPI
Alvin Lebeck	Duke
Ana Del Amo	GE
Andrew Eckford	York University
Aristides Requicha	USC
Boleslaw Szymanski	RPI
Bulent Yener	RPI
Chin-Tser Huang	USC
Chris Carothers	RPI
Chris Dwyer	Duke
Danilo Gligoroski	NTNU
Darren K. Brock	Nantero, Inc.
Dinos Mavroidis	NEU
Dmitri Botvich	WIT
Edmond Jonckheere	USC
Fabrizio Granelli	U Trento
Feng Cheng	U Potsdam
Georgios Lazarou	IEEE
Guillermo Rueda	Intel
Gyorgy Korniss	RPI
Han Yiliang	Xi'an Jiaotong University
Harish Sethu	Drexel
Hong Li	Intel
James Lyke	AFRL
Jian-Qin Liu	NiCT, KARC
Jun Liu	UND
Jure Leskovec	CMU
Kevin Mills	NIST
Khosrow Sohraby	UMKC
Kota Murali	IBM
Lei Liu	Sun Microsystems
Loucas Tsakalakos	GE
Marina Thottan	Lucent
Martha Steenstrup	RCN
Metin Sitti	CMU
Michael Shur	RPI
Murat Yuksel	UNR
Muriel Medard	MIT
Patrick Lincoln	SRI
Paul Bogdan	CMU
Paul Sotiriadis	Johns Hopkins
Rajit Manohar	Cornell
Ralph Droms	Cisco

Sastry Kompella   U Vermont
Satoshi Hiyama   NTT DoCoMo
Seth Copen Goldstein   CMU
Shaker Mousa   ACP
Sherali Zeadally   UDC
Subir Biswas   Michigan State
Suresh Venkatachalaiah   Accenture Technology Solutions
Sylvain Martel   École Polytechnique de Montréal
Tadashi Nakano   UC Irvine
Tong Zhang   RPI
Wei Wang   Ualbany
Wei Yu   TAMU
Yaakov (Kobi) Benenson   Harvard
Yuki Moritani   NTT DoCoMo

Table of Contents

Full Papers

Invited Papers

3D CMOL Crossnet for Neuromorphic Network Applications

Kevin Ryan, Sansiri Tanachutiwat, and Wei Wang

College of Nanoscale Science and Engineering at SUNY Albany,
Albany, 12222 New York, USA
{kryan,stanachutiwat,wwang}@uamail.albany.edu

Abstract. In this work, a novel 3D CMOL crossnet structure is introduced by combining two leading technological concepts for future nanoelectronic neuromorphic networks: CMOL crossnet and 3D integration. By implementing CMOL crossnet into the third dimension, the proposed 3D CMOL crossnet not only maintains the high-speed and high defect-tolerant properties of the CMOS-nano hybrid CMOL hardware system, but also provides efficient fabrication and assembly processes with a much higher density than the original CMOL crossnet. Furthermore, this study focuses on the development of multivalue synapses and efficient communication methods between CMOS and nanodevices. Preliminary results demonstrate that the structure can utilize the advantages of high performance synapses and stable analog CMOS somas in three dimensions. Therefore, the proposed 3D CMOL crossnet structure has a huge potential to become an efficient 3D hardware platform to build neuromorphic networks that are scalable to biological levels.

Keywords: CMOS-Nano Hybrid System, CMOL, Crossnet, Neuromorphic Network, 3D IC.

1 Introduction

For many years, researchers and programmers have been trying to mimic the brains neuromorphic network (NN) in software, and in this task they have been successful [1], [2]. Even though, the complexity of the brains functions have been captured, its efficiency and power at executing these tasks is still an alluring goal. Utilizing today's computer architecture, the software solutions based on a number of enormous supercomputers can not match the brains speed and processing power. In order to provide unparalleled computational efficiency and density, the reconfigurable hard-ware platforms might be a promising solution to mimic the brain's structure [3-5].

Recent studies [6-11] demonstrated that the CMOS-nano hybrid technology can provide an efficient hardware platform to build a family of neuromorphic networks. Based on a hybrid technology, the CMOS parts will implement neural cell bodies (somas) and the nanodevices with reconfigurable capabilities can be used as synapses. The dendrites and axons will be implemented by interconnects or nanowires. This CMOS-nano hybrid neuromorphic network can utilize the advantages of both CMOS and nanodevices.

M. Cheng (Ed.): NanoNet 2008, LNICST 3, pp. 1–5, 2009.

One key challenge of CMOL crossnet is how to efficiently establish communication between CMOS and nanodevices. As shown in Fig. 1a, the CMOL crossnet requires special pins with different heights to connect CMOS and nanodevices, which are difficult to fabricate. Since each nanowire segment requires one pin, millions of these special pins are required, which may not be feasible to build.

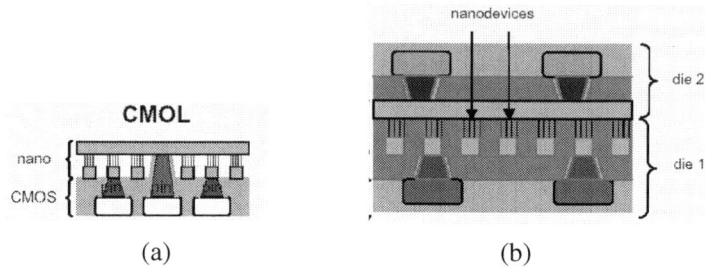

Fig. 1. (a) The CMOL structure with the special pins to connect CMOS and nanodevices [6]. (b) 3D CMOL without the special pin requirement [12].

In order to tackle the challenge of CMOL crossnet, we present our recent research results in developing efficient 3D hybrid neuromorphic network systems. We introduce a 3D CMOL crossnet concept to improve fabrication process, and performance of CMOL crossnet.

2   3D CMOL Crossnet

To ease the fabrication process especially in the area of pin placement and connections, the 3D CMOL crossnet was developed by sandwiching the nanodevices between two separate layers of CMOS (Fig. 1b). Besides the fabrication advantages of this configuration it also doubles the CMOS area and therefore increases the soma density of the architecture. On the other hand, 3D CMOL crossnet will present some design challenges that need to be overcome.

In the brain a soma is the logical unit between the axons and dendrites that controls signal propagation. The soma design presented by Likharev is shown in Fig. 2a and naturally assembles into a matrix type of architecture when put into an array 0. A known benefit of this design is the negative axon-dendrite connection that will always send the opposite signal then its positive counterpart because of the electrical properties of the diode. Since the somas are not able to connect to both layers of nanowires on the crossbar we must take the input and output solely from the vertical or horizontal wires as shown in Fig. 2b. With this arrangement there will be no natural inverse voltage running through the wires, so an inverter has been added before the negative pole of the amplifier to correct the problem.

The final issue to discuss regarding the crossbar structure is the existence of axon-axon and dendrite-dendrite connections. These extra connections which are not found in the brains neuromorphic network are handled quite effectively by the electrical properties of the crossbar. Because of low impedance in the soma's amplifier and

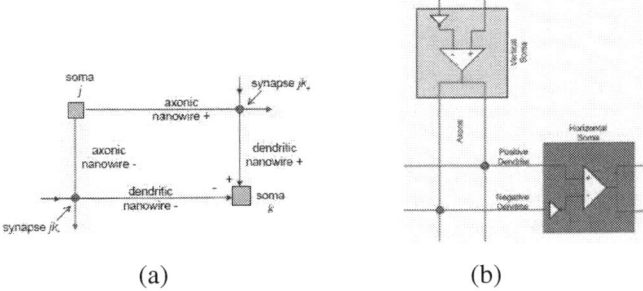

(a)   (b)

Fig. 2. (a) General soma-soma connection as proposed by Likharev (adapted from [7]). (b) Revised 3D CMOL soma-soma connection.

high voltage carried in the axonic nanowires, the axon-axon connections can be disregarded. The dendrite-dendrite connections are not an issue as long as the dendrite voltages are kept lower than the axons hence it will not disrupt the axon's signal already in the wire [8].

Based on the 3D CMOL crossnet structure, we present various designs of neuromorphic networks in this section. These designs tailor the 2D CMOL designs to suit the 3D CMOL crossnet structures. There are two main structures with different variations that can be used to pattern the somas; they are called Alternating FlossBar (Fig. 3a) and Alternating InBar (Fig. 3b). These designs follow an alternating pattern because each soma can only connect to a soma of the opposite type via the synapses. Finally to increase the synapse density we have developed the Cut Alternating FlossBar (Fig. 3c) which makes use of every nanowire junction as a synapse.

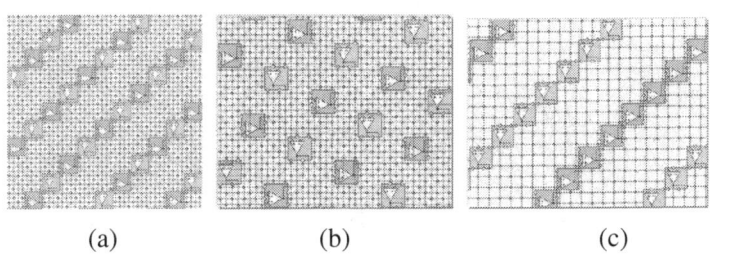

(a)   (b)   (c)

Fig. 3. (a) Alternating FlossBar. (b) Alternating InBar. (c) Cut Alternating FlossBar.

3 Performance Evaluation

In order to evaluate the performance of the proposed 3D CMOL crossnet, we carry out an analysis in terms of area, speed, and power consumption. Note that for simplicity, this analysis does not consider defects and process variations of devices and is not dependent on any specific soma layout.

Table 1 summarizes the estimated implementation results of 2D CMOL crossnet and 3D CMOL crossnet. As shown in Table 1, the proposed 3D CMOL crossnet circuits provide a 2X area improvement over the 2D CMOL crossnet designs with similar operational speed. This is due to 3D CMOL's structure significantly reducing the footprint and logic stages. Also, interconnects and routing of the circuits will be significantly simplified. The 3D CMOL structure can efficiently partition and interconnect lengths and routing complexity to achieve high performance. The average dynamic power of the 3D CMOL designs is a little higher than the original crossnet designs. The short interconnect length of the 3D structure will reduce the dynamic power, but this effect is counteracted by the large capacitance of the nanowire crossbar with its very high density.

A significant advantage of the 3D CMOL crossnet in terms of performance is the density increases compared to the 2D CMOL crossnet. Due to the significant density improvement, this proposed technology would provide huge potential to build future generations of NN applications. However, the challenge of 3D CMOL crossnet lies in the power and thermal management due to its high power density values.

Table 1. Table Performance comparison of 3D CMOL crossnet and 2D CMOL crossnet for spiking and non-spiking models

Non-spiking model	Area (Processing nodes per 858 mm^2)	Speed	Power
CMOL crossnet [11]	1716	50 Hz 140Hz 482 Hz	1.4W
3D CMOL crossnet	3832	<50 Hz <140Hz <482 Hz	<2.8W

Spiking model	Connectivity	Area (Processing nodes per 858 mm^2)	Power
CMOL crossnet [11]	0.1	276	1.8
3D CMOL crossnet	0.1	552	<3.6
CMOL crossnet [11]	0.01	276	2.7
3D CMOL crossnet	0.01	552	<5.4
CMOL crossnet [11]	0.001	276	6.2
3D CMOL crossnet	0.001	552	<12.4

4   Conclusion

In this paper, we have carried out a preliminary study covering the architecture and circuit design of 3D CMOL crossnet. By utilizing high-density nanodevices and high-performance CMOS analog circuits in three dimensions, we can achieve an efficient hardware platform to build neuromorphic network systems. In the future, this CMOS-nano-CMOS one-stack structure can be extended to mutli-stack structures, by bonding several stacks together in a back-to-back manner and connecting them with through-silicon-vias. Such a 3D structure can open up possibilities to develop more complex (hierarchical or modular) neuromorphic systems and brain-like machines. In fact, the human brain essentially is a 3D structure, which can be mimicked by the multi-stack 3D CMOL crossnet.

It is noted that more knowledge of our brain is required to emulate its functionality for advanced intelligent tasks. Recently, neurobiological research has found huge amounts of valuable information related to the brain's structure. Such developments may shed light on future 3D CMOL research. We expect that incorporating the new neuromorphic network research results with the development of 3D CMOL crossnet, may well lead to an innovation or technology breakthrough to construct computers that can match the power of the brain.

Acknowledgments. This work has been supported in part by the AFSTTR and MARCO (via IFC Center). Useful discussions with K. K. Likharev are gratefully acknowledged.

References

1. Varshney, L.R., Sjostrom, P.J., Chklovskii, D.B.: Optimal information storage in noisy synapses under resource constraints. Neuron. 52(9), 409–423 (2006)
2. Wen, Q., Chklovskii, D.B.: Segregation of brain into gray and white matter: a design minimizing coduction delays. PLOS Computational Biology 1(7), 617–639 (2005)
3. Berger, T.W., et al.: Brain-implantable biomimetic electronics as the next era in neural prosthetics. Proceedings of the IEEE 89(7), 993–1012 (2001)
4. Indiveri, G., Chicca, E., Douglas, R.: A VLSI array of low-power spiking neurons and bistable synapses with spike-timing dependent plasticity. IEEE Transactions on Neural Networks 17(1), 211–221 (2006)
5. Vogelstein, R.J., Mallik, U., Vogelstein, J.T., Cauwenberghs, G.: Dynamically reconfigurable silicon array of spiking neurons with conductance-based synapses. IEEE Transactions on Neural Networks 18(1), 253–265 (2007)
6. Strukov, D.B., Likharev, K.K.: A reconfigurable architecture for hybrid CMOS/nanodevice circuits. In: FPGA 2006, pp. 131–140 (Feburary 2006)
7. Türel, Ö., Lee, J.H., Ma, X., Likharev, K.K.: Nanoelectronic neuromorphic networks (crossnets): new results. In: Proc. IJCNN 2004, pp. 389–394 (2004)
8. Türel, Ö.: Devices and circuits for nanoelectronic implementation of artificial neural networks. Ph. D Thesis (2007)
9. Türel, Ö., Lee, J.H., Ma, X., Likharev, K.K.: Neuromorphic Architectures for Nanoelectronic Circuits. Int. J. of Circuit Theory and Applications 32, 277–302 (2004)
10. Lee, J.H., Likharev, K.K.: Defect-Tolerant nanoelectronic pattern classifiers. Int. J. of Circuit Theory and Applications 35, 239–264 (2007)
11. Gao, C., Hammerstrom, D.: Cortical Models Onto CMOL and CMOL-Architectures and Performance/Price. IEEE Trans. Circuit and System I 54(11) (November 2007)

Structural Fault Modelling in Nano Devices

Manoj S. Gaur, Raghavendra Narasimhan, Vijay Laxmi, and Ujjwal Kumar

Malaviya National Institute of Technology, Jaipur-302017, India

Abstract. In this paper we present a model for structural failures in nano-devices. Fault being considered include stuck-at and bridge faults only. This model is an extension of probabilistic model based on Gibbs energy distribution and belief propagation as presented in NANOLAB [1]. Results have been carried out on a 8-bit full adder circuit. Simulation results indicate that probabilistic TMR model represents bridge and stuck-at-1 faults better while deterministic model is more suited for stuck-at-0 faults.

Keywords: Structural fault, stuck-at-0, stuck-at-1, bridge, MRF, TMR.

1 Introduction

Nano-structures are inherently unreliable and this uncertainty stems from low operating energy levels, thermal perturbations/noise, significant quantum effects at nanoscale and high probability of manufacturing defects. Reliable computation requires fault-tolerant architecture and alternate computational model. Triple Modular Redundancy (TMR) and its deviants have been proposed for reliable computation and fault-tolerance [2]. Recently Chen et. al [3] has proposed Markov Random Field and Gibbs' Energy distribution based probabilistic computational model. This probabilistic approach is more reliable computational mode.

Manufacturing defects at nanoscale can lead to permanent structural failures. In addition, thermal perturbations can lead to transient failures. A single NAND gate is not expected to compute reliably at all times. Many fault-tolerant architectures [4] have been reported to deal with this problem. TMR has three similar gates working in parallel and a majority gate to compute output to improve reliability and fault tolerance. Figure 1 illustrates the circuit. Cascade Triple Modular Redundancy (CTMR) has three TMR units of the same type combined with another majority gate to form a 'second-order' TMR unit with a higher reliability. RMR is a generalization of TMR where instead of 3 we have $R = 3, 5, 7, \cdots$ units working in parallel.

2 Probabilistic Model for Nanoscale Computation

Chen [3] proposed a Markov Random Field (MRF) based probabilistic approach for nano-scale computing. This approach adapts to errors as a natural consequence

M. Cheng (Ed.): NanoNet 2008, LNICST 3, pp. 6–10, 2009.

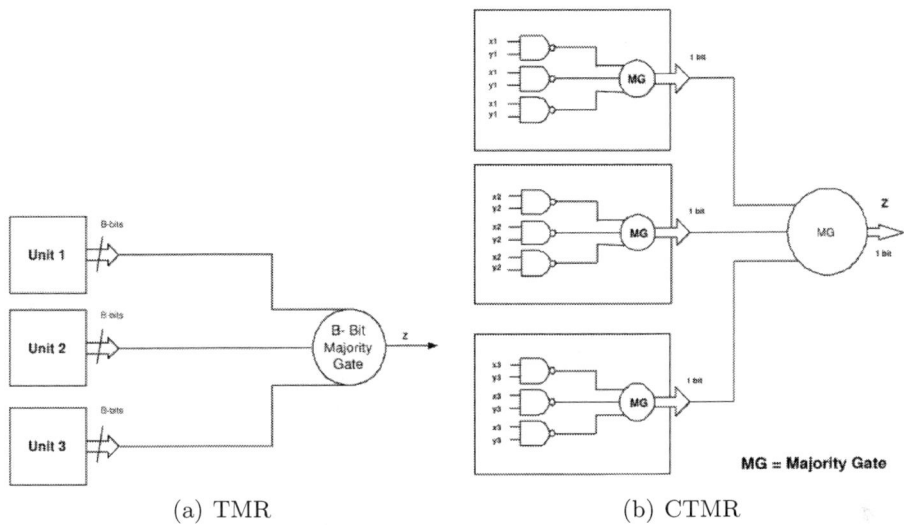

(a) TMR   (b) CTMR

Fig. 1. Fault tolerance through TMR and CTMR

of probability maximization, thereby removing the need to actually detect faults. In MRF model, state of a node depends upon those of its neighbors or clique.

MRF model defines a finite set of random variables, $\Lambda = \{\lambda_1, \lambda_2, ..., \lambda_k\}$. Each variable λ_i has a neighbourhood, $N_i = \{\Lambda - \lambda_i\}$. Energy distribution of any variable depends only on its neighbourhood in the form of a clique. States of nodes in a Boolean network represent variables. In this model, uncertainty and noise handling is through conditional probabilities of energy values with respect to its clique. Gibbs energy distribution is used to model conditional probabilities.

$$P(\lambda_i \mid \{\Lambda - \lambda_i\}) = \frac{1}{Z} e^{\frac{1}{KT} \Sigma_{c\epsilon C} U_c(\lambda)} \tag{1}$$

Here Z is the normalizing constant and, C is the set of cliques. U_c is the clique energy function and depends only on the neighborhood of the nodes. In a logic circuit this clique energy (logic energy) is computed as the summation over the minterms of the valid states ($F_i = 1$). This clique energy definition reinforces that the energy of the invalid logic state is greater than valid state. For a two-input x_0, x_1 NAND gate with output x_2, function $F(x_0, x_1, x_2) = 1$ when $x_2 = (x_0 \wedge x_1)\prime$. The clique energy U (-1 for valid and 0 for invalid state) for NAND gate is

$$U(x_0, x_1, x_2) = -x_2 + 2x_0 x_1 x_2 - x_0 x_1 \tag{2}$$

The probability of the different energy configurations of x_2 is

$$p(x_2) = \frac{1}{Z} \sum_{x_0 \epsilon \{0,1\}} p(x_0) \sum_{x_1 \epsilon \{0,1\}} p(x_1) e^{\frac{-U(x_0, x_1, x_2)}{KT}} \tag{3}$$

2.1   Modeling Noise at Inputs and Interconnects

The probabilistic nondiscrete computing scheme described above can be extended to analyze the impact caused by signal noise at the inputs and interconnects of combinational circuits. For a two input NAND gate, there are three nodes, the inputs x_0 and x_1 and the output x_2. Noise is introduced at the input x_0 as a Gaussian process with mean μ and variance σ^2. The probability distribution of x_2 being in different energy configurations ϵ to $\{0.0, 0.1, 0.2, ..., 0.8, 0.9, 1.0\}$ is

$$p(x_2) = \frac{1}{Z} \int_0^1 \sum_{x_1} e^{\frac{-U}{KT}} \left(\frac{e^{-(x_0-\mu)^2/2\sigma^2}}{\sqrt{2\pi}\sigma} \right) dx_0 \cdot p(x_1) \tag{4}$$

The energy distribution at x_2 if uniform distribution is used to model signal noise is given by the

$$p(x_2) = \frac{1}{Z} \int_0^1 \sum_{x_1 \epsilon \{0,1\}} e^{\frac{-U}{KT}} dx_0 \cdot p(x_1) \tag{5}$$

2.2   NANOLAB

NANOLAB [1] is a tool developed in MATLAB for modelling of logic gates at nanolevel. This model can handle discrete energy distributions at the inputs and interconnects of any specified architectural configuration. These functions work for any generic one, two, or three-input logic gates and can be extended to handle multi-input logic gates and circuits. NANOLAB functions are parameterized and take in as inputs the logic compatibility function and the initial energy distribution for the inputs of a gate. Outcome of these functions is a probability vector indicating the probability of the output node being at different energy levels between [0..1]. These probabilities are calculated over different values of KT so as to analyze thermal effects on the node. The belief propagation algorithm is used to propagate these probability values to the next node of the network. The tool can also calculate entropy values at different nodes of the logic network. It also verifies that, for each logical component of a Boolean network, the valid states have an energy level less than the invalid states. NANOLAB functions can model noise either as uniform or Gaussian distributions or combinations of these, depending on the user specifications. Arbitrary Boolean networks in any redundancy-based defect-tolerant architectural configuration can be analyzed by writing simple MATLAB scripts that use the NANOLAB library functions.

2.3   Fault Models

Fault models are simplifications of phenomena caused by defects on the circuit. The oldest and most common model is the stuck-at fault model. Defects are modelled as a node shorted to either power supply (stuck-at-1) or to ground (stuck-at-0). Bridge fault is an extension of stuck-at fault model wherein two or more links get connected.

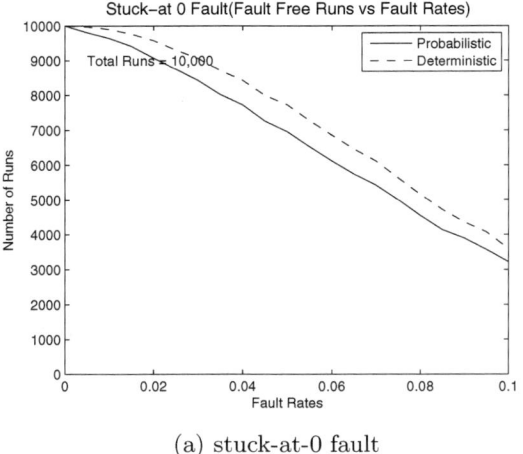

(a) stuck-at-0 fault

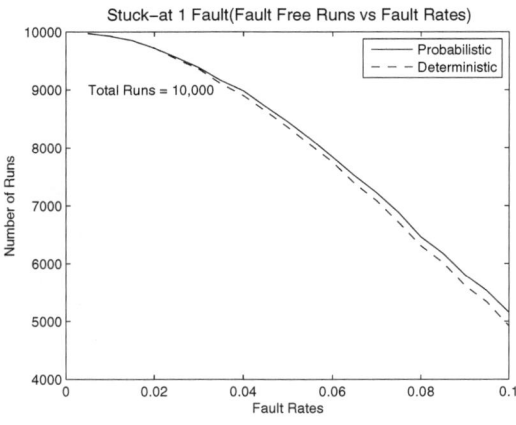

(b) stuck-at-1 fault

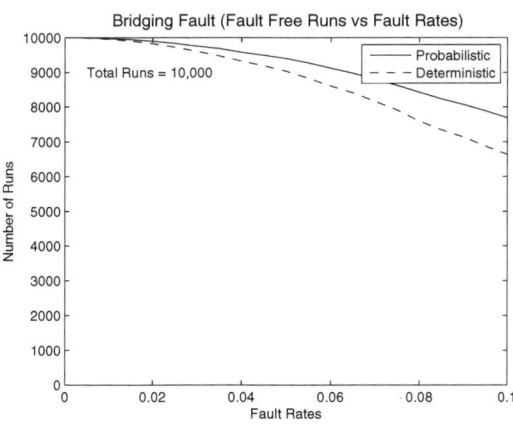

(c) Bridge fault

Fig. 2. Simulation results for structural faults

3   Results

The main circuit that we implemented was a 8-bit full adder circuit. Fault oc-currence was assumed uniform. No gate (link) has more chance of having a fault than other gates (links). At a given location only two links get bridged. Three sets of simulations were run for stuck-at-0, stuck-at-1 and bridge faults. Both traditional deterministic and probabilistic gates were considered in all. Each simulation was repeated $n = 10000$ times with different error rates. For stuck-at fault, error rates ranged from 0.005 to 0.1, with an interval of 0.005. For bridge fault, error rates were varied from 0.5% to 10% with a difference of 0.5% at each step. Output of each simulation was considered 1 (same as expected) or 0 (oth-erwise). In probabilistic case, output equalled logic level with higher probability. Output was labeled invalid, if probability of any logic level was between 0.4 and 0.6. After n simulations number of correct outputs were noted for each case. Figures 2(a), (b) and (c) present results for stuck-at-0, stuck-at-1 and bridge faults respectively.

4   Conclusions

From these graphs, we see that probabilistic gate model gives better result than traditional deterministic one for bridging as well as stuck-at-1 faults. In de-terministic approach the majority gate gives as output which comes maximum number of times. But in probabilistic approach the gate tries to maximize the probability of the output. For example, if the input to the majority gate is [0.8, 0.2],[0.8, 0.2],[0.8, 0.2], output probabilities should be [0.8,0.2]. But in case of probabilistic approach the output comes out to be [0.9 0.1]. However, for stuck-at-0 faults the deterministic approach yields marginally better results than the probabilistic one. In case of NAND gate, the output probability is biased to-wards 1. This is to be expected as its truth table has three ones and only one zero in output. For stuck-at-0 fault, one of the inputs is guaranteed to be 0 im-plying a guaranteed output of 1. In probabilistic approach, some of the states may become invalid if output of two gates in TMR block fall below 0.6.

References

1. Bhaduri, D., Shukla, S.: Nanolab a tool for evaluating reliability of defect-tolerant nanoarchitectures. IEEE Transactions on Nanotechnology 4(4) (July 2005)
2. von Neumann, J. (ed.): Probabilistic Logic and the Synthesis of Reliable Organisms from Unreliable Components. Princeton University Press, Princeton (1956)
3. Chen, J., Mundy, J., Bai, Y., Chan, S.M.C., Petrica, P., Bahar, R.I.: A probabilis-tic approach to nano-computing. Technical report, Division of Engineering, Brown University, RI 02912, USA
4. Nikolic, K., Sadek, A., Forshaw, M.: Architecture for reliable computing with unre-liable nanodevices. IEEE-NANO M2.1 Nano-Devices (II), 254–259 (2001)

Proposal for Memristors in Signal Processing

B. Mouttet

George Mason University
4400 University Drive, Fairfax, Va. 22030
bmouttet@gmu.edu

Abstract. Recently researchers at Hewlett-Packard have announced the discovery of a new material having resistance switching characteristics and which has been characterized as a fourth fundamental circuit component called the "memristor"[1]. It is proposed to combine such memristors with operational amplifier circuitry and fixed resistor elements so as to form a programmable signal processor capable of selective transmission and multiplexing of multiple signals for applications in communications and programmable drive waveform control.

Description of Device

While the recent revelation by Hewlett-Packard of bi-layer titanium oxide as a candidate material for the "missing memristor" initially speculated by Leon Chua [2] is an interesting and important development, it is not entirely unprecedented. Similar resistance variability effects in thin film oxides have been studied using perovskite [3] and similar materials. However, this research has mostly been limited to applications in non-volatile memory in which the switching resistance acts to store binary data in the form of high or low resistance states. Such materials may also have use in signal processing and control applications.

Fig. 1 illustrates an idealized approximation for the behavior of memristance material in which an applied voltage greater than some positive threshold voltage V_{L1} initiates a memristance region exhibiting variation from a high resistance level R_H to a low resistance level R_L as voltage is increased. Similarly, resistance may be converted from a low level back to a high level by a reversed polarity voltage in the region between $-V_{L2}$ and $-V_{H2}$. In the small signal region between $-V_{L2}$ and V_{L1} the material is either at a high or low resistance level depending on the history of voltage application to the material.

Fig. 2 illustrates an array of memristors M_1-M_4 having inputs connected to fixed resistors R_1-R_4 and outputs connected to a common inverting input of an operational amplifier having a feedback resistance R_F. Operational amplifiers in such a configuration exhibit the well known effect of summing the input signals based on a weighting determined by a ratio of the feedback resistance and the input resistances. However, the inclusion of memristors allows controllability in selecting which signals to sum and, if tuned within the memristance region, the weighting values of each input signal may also be adjusted. By using harmonic sinusoidal signals or step

M. Cheng (Ed.): NanoNet 2008, LNICST 3, pp. 11–13, 2009.

signals with relative time delays as the voltage signal inputs this circuit configuration may be used for programmable waveform generation useful for automatic control systems. Programmable signal mixing and modulation may also be facilitated using this configuration by switching memristor values during communication applications such as frequency hopping.

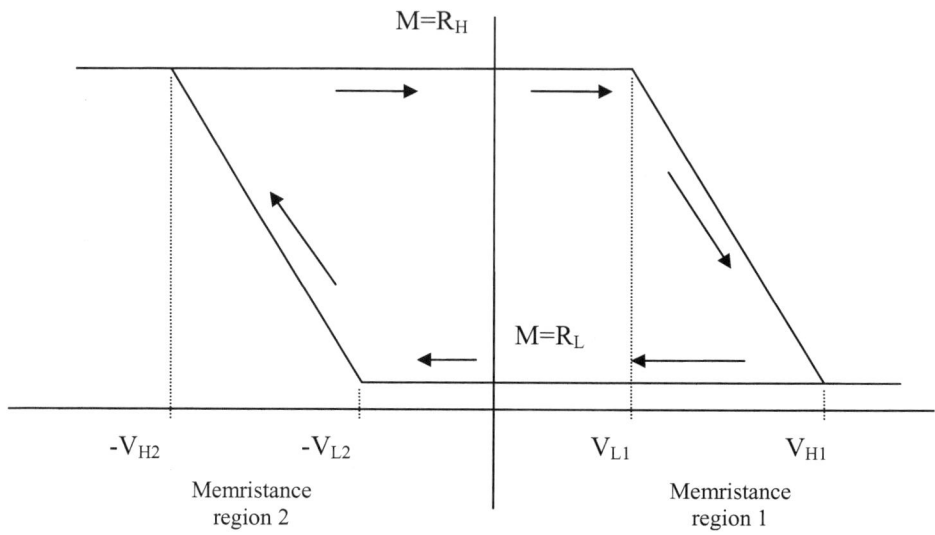

Fig. 1. Idealized hysteresis model of resistance vs. voltage for memristance switch

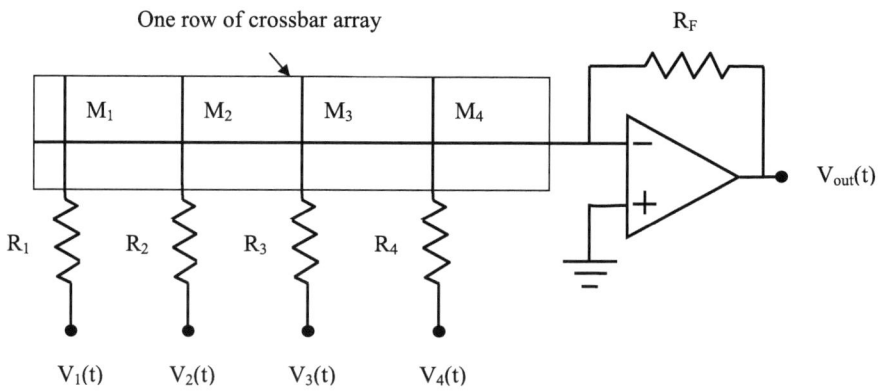

$$V_{out}(t) = -\sum_i V_i(t) * R_F /(R_i + M_i) \qquad M_i = R_H \text{ or } R_L \text{ for } -V_{L2}<V_i(t)<V_{L1}$$

By setting the fixed resistors $R_F \ll R_H$ and $R_i = R_F - R_L$ the output signal may be approximated by:

$$V_{out}(t) \cong -\sum_i V_i(t) * \delta(M_i), \ 0 \le \delta(M_i) \le 1 \text{ depending on the memristance states.}$$

Fig. 2. Summing amplifier configured with memristance elements

References

1. Strukov, D.B., Snider, G.S., Stewart, D.R., Williams, R.S.: The Missing Memristor Found. Nature 453(7191), 42–43 (2008)
2. Chua, L.: Memristor-The missing circuit element. IEEE Transactions on Circuits Theory 18(5), 507–519 (1971)
3. Beck, A., Bednorz, J.G., Gerber, C., Rossel, C., Widmer, D.: Reproducible switching effects in thin oxide films for memory applications. Applied Physics Letters 77(1), 139 (2000)

Normal and Reverse Temperature Dependence in Variation-Tolerant Nanoscale Systems with High-k Dielectrics and Metal Gates

David Wolpert and Paul Ampadu

Electrical and Computer Engineering Department
University of Rochester, Rochester, NY 14627, USA
{wolpert,ampadu}@ece.rochester.edu

Abstract. The delay dependence on temperature reverses at increasingly larger supply voltages as technology scales into the nanometer regime, causing delay to decrease as temperature increases. This reversal can be problematic for variation-tolerant systems using critical path replicas to determine delay guardbands, as delay may no longer indicate when the system is in danger of thermal runaway. Adaptive voltage scaling, commonly used in variation-tolerant systems, further complicates the temperature impact, as the range of voltages may intersect both temperature regions. In this paper, it is shown that use of high-k dielectrics and metal gates increases the supply voltage where this reversal occurs by 40% compared to low-k, poly gate technologies. 45, 32, and 22 nm models are examined, and the reversal voltage is shown to approach 90% of nominal voltage at 22 nm, making the effect important even for non-adaptive designs. Techniques to account for these complex temperature dependencies are proposed to ensure functionality under all conditions.

Keywords: Reverse temperature dependence, variation-tolerant, high-k dielectric, metal gate.

1 Introduction

Operating temperature affects device delay by altering mobility (μ) and threshold voltage (V_T), according to [1]

$$\mu(T) = \mu_0 (T/T_0)^{\alpha_\mu} \tag{1}$$

and

$$V_T(T) = V_{T0} + \alpha_{V_T}(T - T_0). \tag{2}$$

where T_0 is the nominal temperature (generally 300 K), μ_0 is mobility at T_0, α_μ is an empirical parameter referred to as the mobility temperature exponent, V_{T0} is nominal threshold voltage, and $\alpha_{V_T} = \partial V_T / \partial T$ is another empirical parameter named the threshold voltage temperature coefficient.

M. Cheng (Ed.): NanoNet 2008, LNICST 3, pp. 14–18, 2009.

V_T, μ, and nominal supply voltage (V_{DD}) are all technology dependent parameters, with predicted values available down to the 22 nm node [2,3]. Use of high-k dielectrics and metal gates to alleviate nanoscale gate leakage problems also alters V_T and μ [4,5]. The combination of these changes makes it difficult to determine the effect of temperature on device performance. Two temperature regions exist: normal temperature dependence, where current decreases as temperature increases, and reverse temperature dependence [6,7], where current increases as temperature increases.

These parameters are further complicated by environmental requirements (military specifications call for a range of -55°C to 125°C) and intra-die temperature variation (shown to exceed 25°C [8]). To account for the wide range of conditions, as well as process and voltage variations, variation-tolerant adaptive systems have been used to guarantee functionality by adjusting operating voltages and frequencies [9,10,11]; however, these systems with multiple voltage modes make the above-mentioned temperature effects even more difficult to determine.

The remainder of this paper will be organized as follows: In Section 2, the impact of high-k dielectrics and metal gates on temperature behavior will be examined, and the changing impact of temperature across 45, 32, and 22 nm technologies will be shown. In Section 3, the effects of these complex temperature dependencies on variation-tolerant systems are explained, and techniques for considering these dependencies are discussed. Conclusions are presented in Section 4.

2   Temperature Impact on Current and Delay

For large gate overdrives ($V_{GS} - V_T > V_{ti}$, where V_{ti} is an empirical parameter referred to as the temperature insensitive voltage [7]), the temperature dependence of a device is dominated by the dependence of μ, while for small gate overdrives ($V_{GS} - V_T < V_{ti}$), small changes in V_T can cause large changes in current, resulting in a temperature dependence dominated by V_T. The normal temperature dependence occurs when the μ dependence dominates, while the reverse temperature dependence occurs when the V_T dependence dominates. Further examination of these effects in low-k dielectric, polysilicon gate devices is available in [6,7].

In nanoscale devices, high-k dielectrics and metal gates have been introduced to reduce gate leakage due to thinning gate oxides and reduce the depletion effects of polysilicon gates [4,5]; unfortunately, these techniques have the effect of dramatically increasing the temperature dependence on V_T. The extent of this effect is shown in Fig. 1, which compares 45 nm predictive technology models [2] of both low-k/poly gate (dashed line) and high-k/metal gate (solid line) devices. Each line in Fig. 1 shows the change in delay of an inverter ($\beta = 2$) from -55°C to 125°C. For example, at 0.62 V, the high-k/metal gate inverter delay does not change at all from -55°C to 125°C, resulting in the 0.62 V point occurring on the 0% line. This 0% intersect on each curve represents V_{ti}. As shown, the high-k/metal gate devices result in a 40% increase in V_{ti} compared to the low-k, polysilicon gate devices. The normal temperature dependence region is below the 0% line, and the reverse dependence region is above the 0% line.

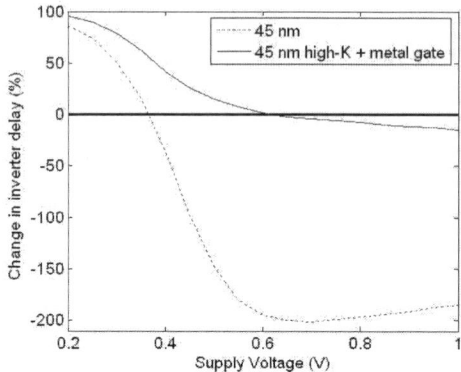

Fig. 1. Effect of high-k dielectric and metal gate on temperature dependence

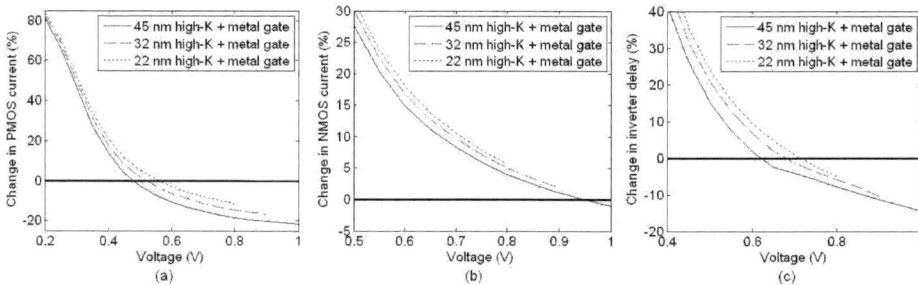

Fig. 2. Changes in (a) PMOS current, (b) NMOS current, and (c) inverter delay over the -55°C to 125°C temperature range

Fig. 2(a) shows the change in PMOS device current from -55°C to 125°C at the 45, 32, and 22 nm technology nodes. Nominal voltage at each node is indicated by the rightmost point on each curve, equal to 1 V, 0.9 V, and 0.8 V, at 45, 32, and 22 nm, respectively [2,3]. As shown, V_{ti} of the PMOS devices steadily increases as technology scales down by about 400 mV (20%) per node, with V_{ti} at 22 nm equal to 0.56 V. The NMOS device response, shown in Fig. 2(b), is quite different, with nominal voltages at the 32 and 22 nm nodes already in the reverse temperature dependence region. The PMOS and NMOS devices are combined into an inverter in Fig. 2(c), with $\beta = 2$ to represent FO4 minimum delay sizing. As shown, V_{ti} in the inverter approaches 90% of nominal voltage in the 22 nm node. As β increases, the stronger PMOS effect further increases V_{ti}. Thus, these complex temperature effects will require attention in nanoscale systems even at nominal voltages.

3   Variation-Tolerant Systems with Complex Temperature Dependences

Reverse temperature dependence at near nominal voltages complicates variation-tolerant system design, which uses multiple supply voltages to adjust for changes in process, voltage, and temperature. The additional complexity needed to account for both normal and reverse temperature dependence depends on the design time information. If the system can be fully characterized at design time, then this can be solved by updating the voltage and frequency look-up table entries [10] to ensure that the system adapts in the correct direction given a change in temperature. For example, whereas a low-voltage system would generally reduce the frequency as temperature increases, in the reverse dependence region the system would have to reduce the frequency when temperature decreases.

If the temperature regions in some voltage modes are not known at design time, whether due to tool limitations, process variations, or unknown IR drops, they must be determined at runtime. If the system is known to be in the normal temperature dependence region at nominal voltage, then a fixed nominal-voltage ring oscillator can be compared to the critical path replica (whose supply voltage changes with the rest of the system to track the delay at each voltage mode). By comparing the critical path replica delay with the ring oscillator frequency at two different temperatures, the temperature region of each voltage mode can be determined at manufacture time and stored. To include aging effects, this test can be performed at runtime as needed.

If the temperature dependence is not known for every voltage mode, then there are two options for ensuring variation-tolerance. One option is to design the system with large enough guardbands that it can operate correctly over the entire temperature range regardless of the dependence, though this will result in a large reduction in delay performance. Another option is to use a temperature sensor with a poly resistor, which would consume large area and power but avoid the nonlinear effects of mobility and threshold voltage, resulting in a stable reference exhibiting normal temperature dependence regardless of variations.

An important issue with all of the approaches mentioned in this section is that high temperatures in the reverse dependence region are no longer self-limiting: In the normal dependence region, temperatures are unable to increase to dangerous levels because the delay would become so large that the system would be forced to throttle the frequency, reducing the energy and therefore the temperature. In the reverse dependence region, the circuits will continue to speed up as temperature increases, with no such delay problem. This could potentially result in race conditions, or even more concerning, the higher temperatures could result in thermal runaway due to the exponential temperature dependence of leakage current [12], which will already be dominating the total power consumption in the nanoscale regime [13].

4   Conclusion

While the complexities of temperature dependence were previously only an issue in ultra-low voltage design, the combination of high-k dielectrics, metal gates, and nanoscale parameters mean they will require attention even in nominal voltage systems. The reversal of temperature dependence occurs at 90% of the nominal supply voltage at 22 nm, and potentially even higher voltages depending on device ratios. After considering voltage and process variation effects, unknown temperature dependences may affect any system, whether or not it uses adaptive voltage controls. Techniques for avoiding delay failure and thermal runaway as a result of these complex dependencies will become more and more important as technologies stretch further into the nanoscale regime.

References

1. Filanovsky, I.M., Allam, A.: Mutual compensation of mobility and threshold voltage temperature effects with applications in CMOS circuits. IEEE Trans. Circuits and Systems I: Fundamental Theory and Applications 48, 876–884 (2001)
2. Zhao, W., Cao, Y.: New generation of Predictive Technology Model for sub-45nm early design exploration. IEEE Trans. on Electron Devices 53, 2816–2823 (2006)
3. Zhao, W.: Personal communication (2008)
4. Guillaumot, B., et al.: 75nm damascene metal gate and high-k integration for advanced CMOS devices. In: International Electron Devices Meeting, pp. 355–358 (2002)
5. Cheng, B., et al.: The impact of high-k gate dielectrics and metal gate electrodes on sub-100 nm MOSFET's. IEEE Trans. Electron Devices 46, 1537–1544 (1999)
6. Park, C., et al.: Reversal of temperature dependence of integrated circuits operating at very low voltages. In: International Electron Devices Meeting, pp. 71–74 (1995)
7. Bellaouar, A., Fridi, A., Elmasry, M.I., Itoh, K.: Supply voltage scaling for temperature-insensitive CMOS circuit operation. IEEE Trans. Circuits and Systems II: Analog and Digital Signal Processing 45, 415–417 (1998)
8. Sato, T., Ichimiya, J., Ono, N., Hachiya, K., Hashimoto, M.: On-chip thermal gradient analysis and temperature attening for SoC design. IEICE Trans. Fundamentals E88-A, 3382–3389 (2005)
9. Martin, S., Flautner, K., Mudge, T., Blaauw, D.: Combined dynamic voltage scaling and adaptive body biasing for lower power microprocessors under dynamic workloads. In: IEEE/ACM Int. Conf. on Computer Aided Design, pp. 721–725 (2002)
10. Tschanz, J., et al.: Adaptive frequency and biasing techniques for tolerance to dynamic temperature-voltage variations and aging. In: IEEE Int. Solid-State Circuits Conf., pp. 292–293 (2007)
11. Elgebaly, M., Sachdev, M.: Variation-Aware Adaptive Voltage Scaling System. IEEE Trans. Very Large Scale Integration (VLSI) Systems 15, 560–571 (2007)
12. Lee, C.-C., de Groot, J.: On the thermal stability margins of high-leakage current packaged devices. In: 8th Electronics Packaging Technology Conference, pp. 487–491 (2006)
13. Kim, N.S., et al.: Leakage current: Moore's law meets static power. IEEE Computer 36, 68–75 (2003)

NEMS Capacitive Sensors for Highly Sensitive, Label-Free Nucleic-Acid Analysis

Manu Sebastian Mannoor[1], Teena James[1], Dentcho V. Ivanov[1], Les Beadling[2], and William Braunlin[2]

[1] New Jersey Institute of Technology, Newark NJ 07102, USA
msm28@njit.edu
[2] Rational Affinity Devices LLC, Newark, New Jersey 07103, USA

Abstract. A highly sensitive NEMS capacitive sensor with electrode separation in the order of Debye length is fabricated for label free DNA analysis. The use of nano-scale electrode separation provides better insight in to the target-probe interaction which was not previously attainable with macro or even micro scale devices. As the double layers from both the capacitive electrodes merge together and occupy a major fraction of the capacitive volume, the contribution from bulk sample resistance and noises due to electrode polarization effects are eliminated. The dielectric properties during hybridization reaction were measured using 10-mer nucleotide sequences. A 45-50% change in relative permittivity (capacitance) was observed due to DNA hybridization at 10Hz. Capacitive sensors with 30nm electrode separation were fabricated using standard silicon micro/nano technology and show promise for future electronic DNA arrays and high throughput screening of nucleic acid samples.

Keywords: Micro/Nanofabrication, Capacitive Sensor, Biosensors, DNA detection.

1 Introduction

Over a decade of rapid advances in Micro and Nano fabrication technologies has opened up enormous possibilities across various fields of science and technology. The integration of microelectronics technology with molecular biology is having a transforming impact in the development of biosensors with potential applications in future drug and diagositic development. By the use of miniaturization techniques, the sensing elements or at least parts of them are now getting shrunk down to the same order of dimension as the biomolecules being sensed, resulting in the improvement of many attributes of the molecular detection processes. These nano-scale sensors offer solutions to many problems suffered by conventional signal transduction mechanisms, thereby improving detection sensitivity immensely.

Biosensing, in general involves the detection or quantification of specific biochemical agents such as a particular DNA sequence or protein, using a biorecognition layer for specificity, which is usually immobilized on a transducer surface.

M. Cheng (Ed.): NanoNet 2008, LNICST 3, pp. 19–25, 2009.

Several physiochemical signal transduction mechanisms such as optical, magnetic, electrochemical and piezoelectric have been demonstrated over the past decades for the generation of a physical signal from the binding/hybridization events[1][2][3][4]. The dielectric spectroscopic measurements conducted on micro fabricated capacitive structures gained special attraction due to their label free operation and absence of any mechanical motion [5].

Although, several examples of capacitive biosensor have been reported in the literature, many physical and electrochemical properties of these structures and the measurement methods used have significantly limited their commercial full-scale development as a biosensor. The existence of electrode polarization effect and noises from solution conductance limited the earlier dielectric spectroscopic measurements to high frequencies only, which in turn limited its sensitivity to biomolecular interactions, as the applied excitation signals were too fast for the charged macromolecules to respond [6][7][8]. The series parasitic impedance from electrode polarization effect masked the dielectric changes occurring due to biomolecular interactions at low frequencies (<1 kHz) and the proposed methods for minimizing this effect were not compatible with bio sensing applications [9][10].

In an attempt to address the above mentioned challenges, we report a NEMS capacitive sensor with electrode separation in the order of Debye length. The use of a 30nm electrode separation provides better insight into the molecular interactions, which was not previously attainable with macro or even micro scale devices. As the double layers from both the capacitive electrodes merge together and occupy a major fraction of the capacitive volume, the contribution from bulk sample resistance in the measured impedance will be eliminated. The interaction between the electrical double layers due to the space confinement decreases the potential drop across the electrode spacing and allows dielectric measurements at low frequency.

2   Experimental

2.1   Device Fabrication

The most critical parameter for enhancing sensitivity by eliminating the electrode polarization effect is the nanometer separation between the capacitive electrodes. The desired separation of less than 50nm is difficult to achieve with conventional lithographic techniques [11]. To overcome the resolution limit, we have used a sacrificial layer process where the thickness of the SiO_2 spacer layer determines the electrode separation. The process steps are schematically indicated in Fig 1. In the first process step 500nm thick Silicon Nitride is deposited on the single side polished <100> Si wafers after which a $1\mu m$ thick photo resist spacers are patterned to act as the sacrificial layer for the formation of the first set of Au electrodes (a). Gold electrodes are deposited using E-beam evaporation under ultra high vacuum conditions. The selective removal of the photo resist sacrificial layer defines the first set of Au electrodes (b). In the next step a very thin and uniform layer of SiO_2 is deposited using Plasma Enhanced Chemical

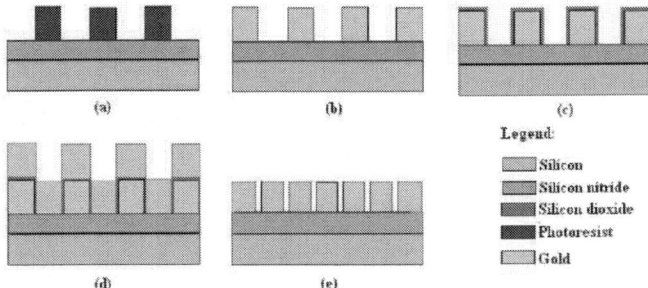

Fig. 1. Schematics of the fabrication process flow. a) photo resist spacers are patterned b) gold electrodes formed by sacrificial method c) deposition of SiO$_2$ for nanometer spacing d) deposition of gold e) SiO$_2$ spacer removed.

Vapor Deposition (PECVD), to form the nanometer spacers between the electrodes (c). Followed by the patterning of SiO$_2$ sacrificial layer, a second layer of gold metallization of 1μm is done using E-beam evaporation (d). The Au electrodes were planarized by CMP and finally the SiO$_2$ spacer film between the gold electrodes is selectively etched off using HF (e). The deposited Silicon Nitride layer will act as etch stop of this etching process and also serves as an isolator between the gold electrodes and the Si wafer. Here the use of deposited oxide thin film to define the separation between gold electrodes allows the fabrication of capacitive structures with electrode separations lower than the resolution limit of optical or e-beam lithography.

Detection Scheme. The detection is based on the changes in sensor capacitance due to the variation in dielectric properties of the Debye layer resulting from the biomolecular interactions. The Debye capacitance (double layer

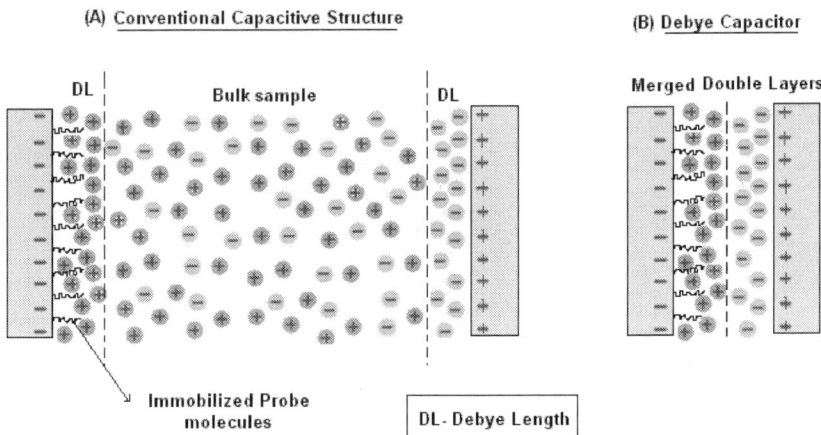

Fig. 2. Comparison between conventional capacitive sensor and Debye Capacitive Sensor

capacitance) formed by the accumulation of counter ions near the electrode surface is highly sensitive to the changes in the dielectric and charge environment in
the electrode/electrolyte interface. The characteristic length of this diffuse double layer of charges (Debye layer) is given by the Debye-Hckel parameter and
is generally called the Debye length. The calculated Debye length for the buffer
solution used in our experiment is 76 nm. The use of capacitive element with
electrode separation of 30nm results in the overlapping of the Debye layers of the
two electrodes. Fig 2 shows a comparison between the working of a conventional
capacitive sensor and the Debye capacitive sensor.

Results and Discussion. The dielectric properties were investigated over a
frequency range of 10Hz to 100 kHz, with 0V DC bias and 20mV AC signals
using an SR 785, 2 channel dynamic signal analyzer. A Lab View program is used
to collect and record data through a GPIB interface. The electrical contacts and
the functioning of the entire system including the capacitive element are verified
by measuring the dielectric spectrum with air and De Ionized water in between
the electrodes. The relative permittivity values of various concentrations of buffer
solutions are measured to verify the properties of the Debye layer. Figure 3 shows
the relative permittivity values obtained for the different buffer concentrations
and are seen to increases with increasing concentration. This can be explained
by the fact that the Debye length decreases with increasing concentration which
results in an increase in potential drop across the sensor.

DNA oligonucleotides used for the experiments were purchased from IDT (Integrated DNA Technologies). Other chemicals including the buffer solution were
purchased from Sigma-Aldrich. Single stranded probe DNA sequences
premodified by the thiol linker(5'-CACGTAGCAG/3 Thio MC3-D/-3') were
immobilized on the gold electrodes using a concentration of $10\mu M$ in 0.05xSSC
buffer (7.5mM Sodium Chloride + 0.75mM Sodium Citrate). By taking advantage of the high affinity of sulphur atoms to gold substrate, the DNA molecules
with thiol end groups are chemically assembled onto the gold surface from the
solution. The oligomer chains are thus tethered to the gold substrate at one end

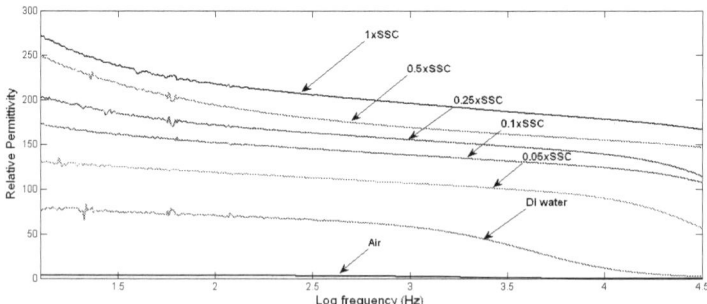

Fig. 3. Relative permittivity as a function of frequency for various concentrations of
buffer solutions.The measured permittivity increased as the ionic strengths are increased.

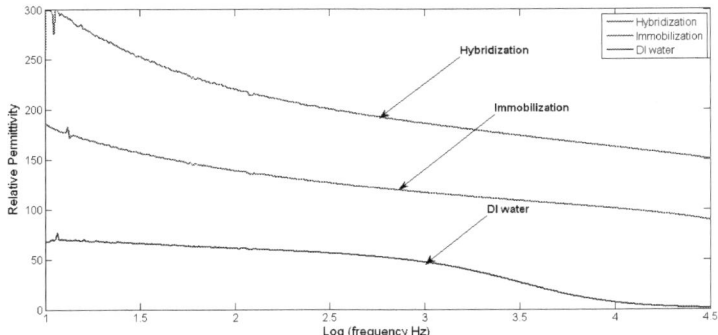

Fig. 4. Dielectric Spectra after hybridization of (10μM) complementary target with the immobilized probe sequence

and the rest of the chain stays fully extended at an angle of approx. 30 degree from the surface. The Van der Waals forces between adjacent chains helps to order the oligomers parallel to each other .The <111> crystal orientation of gold which is obtained by thin film deposition is found to give excellent result for the formation of self assembled monolayers (SAM). Mercapto hexanol (HS- $(CH_2)_6$ OH) SAM layers were immobilized in between the DNA strands in order to passivate the vacant spaces. Prior to immobilization procedure the structure was cleansed using acetone, isopropanol and deionized water. The substrates with immobilized oligomers were then allowed to interact with $0.1\mu M$ to $10\mu M$ concentration of complementary oligomers (5'-CTG CTA CGT G-3') over a short period of time. After incubation, the substrates were rinsed with deionized water to remove the nonspecifically bound target molecules.

The rel.permittivity changes after probe immobilization and target hybridization are shown in Fig 4a. The increased potential drop across the electrical

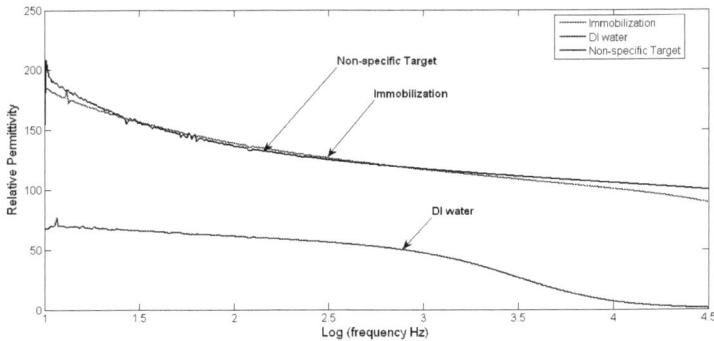

Fig. 5. Dielectric Spectra after the interaction of non-complementary target with the immobilized probe sequence. The slight variation in rel.permittivity value indicates the non-specifically bound oligomers.

double layer due to the additional layer of hybridized oligomers is reflected as the increase in the overall sensor permittivity (capacitance).

As a control experiment, a non complementary target (5'-ATG GCC CTG T-3') solution in the same concentration as the complementary target solution is allowed to interact with the immobilized probe layer. Fig 5 shows negligible change in dielectric property upon exposure to the non complementary sequence. This supports the relationship between capacitance change and specific nucleotide interaction.

3 Conclusion

Capacitive biosensors with electrode separation in the order of electrical double layer width were designed and fabricated using SiO_2 sacrificial layer techniques. The nano scale electrode space confinement is shown to eliminate noises from electrode polarization effect and solution conductivity, permitting the dielectric spectroscopic measurements at low frequencies. DNA hybridization experiments with complementary and non complementary target sequences were performed and a 45-50% change in sensor permittivity (capacitance) was observed after the hybridization of the immobilized probe with the complementary oligomer sequences. Work is presently being carried out in order to optimize the detection mechanism and improve the sensitivity and selectivity of the sensor. Ideally, Single Nucleotide Polymorphism (SNP) could be detected with the current geometries of the device. The improved sensitivity demonstrated by the Debye Capacitive sensor combined with its use of reduced sample volume and low fabrication cost makes it promising for applications such as point of care diagnostics and biowarfare agent detection.

Acknowledgement

The fabrication of the sensors was performed in the Micro Fabrication Center (MFC), New Jersey Institute of technology (NJIT). We express our gratitude to Dr. Rajendra K. Jarwal of MFC for assistance in the fabrication procedures and for many key discussions.

References

1. Rowe-Taitt, C.A., Golden, J.P., Feldstein, M.J., Cras, J.J., Hoffman, K.E., Ligler, F.S.: Array biosensor for detection of biohazards. Biosens Bioelectron 14, 785–794 (2000)
2. Miller, M.M., Sheehan, P.E., Edelstein, R.L., Tamanaha, C.R., Zhong, L., Bounnak, S., Whitman, L.J., Colton, R.J.: A DNA array sensor utilizing magnetic microbeads and magnetoelectronic detection. Journal of Magnetism and Magnetic Materials 225, 138–144 (2001)
3. Rudel, U., Geschke, O., Cammann, K.: Entrapment of Enzymes in Electropolymers for Biosensors and Graphite Felt Based Flow-Through Enzyme Reactors. Electroanalysis 8, 1135–1139 (1996)

4. Vaughan, R.D., O'Sullivan, C.K., Guilbault, G.G.: Sulfur based self-assembled monolayers (SAM's) on piezoelectric crystals for immunosensor development. Fresenius' Journal of Analytical Chemistry 364, 54–57 (1999)
5. Daniels, J.S., Pourmand, N.: Label-free impedance biosensors: Opportunities and challenges. Electroanalysis 19, 1239–1257 (2007)
6. Lee, J.S., Choi, Y.K., Ha, D., King, T.J., Bokor, J.: Low-frequency noise characteristics in p-channel FinFETs. IEEE Electron Device Letters 23, 722–724 (2002)
7. Oleinikova, A., Sasisanker, P., Weingartner, H.: What can really be learned from dielectric spectroscopy of protein solutions? A case study of ribonuclease A. Journal of Physical Chemistry B 108, 8467–8474 (2004)
8. Sanabria, H., Miller, J.H.: Relaxation processes due to the electrode-electrolyte interface in ionic solutions. Physical Review E - Statistical, Nonlinear, and Soft Matter Physics 74 (2006)
9. Fricke, H., Curtis, H.J.: The dielectric properties of water-dielectric interphases. Journal of Physical Chemistry 41, 729–745 (1937)
10. Schwan, H.P.: Electrical properties of tissue and cell suspensions. Advances in biological and medical physics (1957)
11. Nuzzo, R.G., Allara, D.L.: Adsorption of bifunctional organic disulfides on gold surfaces. Journal of the American Chemical Society 105, 4481–4483 (1983)

Impact of Process Variation in Fault-Resilient Streaming Nanoprocessors

Michael Leuchtenburg, Pritish Narayanan, Teng Wang,
and Csaba Andras Moritz

UMass Amherst, Amherst MA 01002, USA
{mleuchte,pnarayan,twang,andras}@ecs.umass.edu

Abstract. We show results from ongoing work studying the interaction of process variation and built-in fault resilience intended to handle defects. We find that built-in fault resilience decreases the negative effects of process variation on a streaming nanoprocessor design.

Keywords: nanoscale processor, process variation, defect tolerance.

1 Background

All nanoscale circuit architectures include a defect-tolerance mechanism to handle the high defect rates expected with bottom-up manufacturing techniques. This is required to achieve acceptable yield, as has been extensively studied. One such mechanism is built-in fault resilience, where logic is added to allow the circuit to function even if some of its components are not functioning correctly. Some proposed systems use reconfiguration to work around defects, but built-in fault resilience may be required to tolerate faults from other sources.

Another major issue with nanoscale circuits, as with CMOS circuits today, is process variation. Process variation leads to large variation in delay, which may cause incorrect function of the system due to missed deadlines and also makes it difficult to achieve high performance.

We are using WISP-0, a simple 5-stage streaming processor design, based on the NASIC nanoscale fabric architecture, to explore the impact of process variation on circuits with built-in fault resilience. NASIC is a tiled 2-D grid-based circuit fabric using a dynamic circuit style. For more information on NASIC and WISP-0, please see [1].

2 Simulation Results

We have used results from the literature [2] [3] [4] [5] to implement a timing model for NASIC. We use the WISP-0 design in our simulations so as to gauge the effects on a processor. In order to capture the typical behavior, we use the Monte Carlo method, picking the parameter values used for each wire and transistor from a distribution. We use a Gaussian distribution with a variation of

M. Cheng (Ed.): NanoNet 2008, LNICST 3, pp. 26–27, 2009.
© ICST Institute for Computer Sciences, Social Informatics and Telecommunications Engineering 2009

$3\sigma = 60\%$ for each parameter, which is higher than reported by nanoscale device and materials researchers. Some of the parameters varied are the diameter of the nanowires, the resistivity of the wires, the contact resistance with the microscale wires, the pitch of the grid, and the thickness of the gate oxide.

For each set of parameter values, the minimum delay at which the processor can achieve correct results is determined through simulation. Circuits which cannot operate correctly at any speed are not considered in the delay statistics.

The type of fault resilience used is integrated error correction (IEC) based on Hamming codes combined with 2-way redundancy (IEC). Defects considered are both stuck-on and stuck-off transistors, with the proportions assumed to be 90% stuck-on due to the structure and devices used in NASIC. For more details about the fault model and fault resilience technique, see [1].

Table 1. Delay per cycle with and without process variation (in picoseconds)

No Variation	With Variation ($3\sigma = 60\%$)		
No defects	No defects	5% defects	10% defects
182.15	178.47	182.4	184.41

In a traditional circuit, we would expect to see the typical case be slower than what could be achieved without process variation. However, as can be seen in Table 1, the mean delay for WISP-0 with fault resilience is instead decreased. This decrease in delay is due to the IEC handling timing faults, thus allowing the speed to be pushed further. Since only a limited number of faults can be tolerated by each stage, as the defect rate goes up, the delay increases as well.

Circuits with built-in fault resilience clearly also have some tolerance for process variation. We are working on techniques to further decrease the impact of process variation on performance, especially in the presence of defects, and to increase the overall performance of NASIC circuits.

References

1. Moritz, C.A., Wang, T., Narayanan, P., Leuchtenburg, M., Guo, Y., Dezan, C., Ben-Naser, M.: Fault-Tolerant Nanoscale Processors on Semiconductor Nanowire Grids. IEEE Transactions on Circuits and Systems I: Fundamental Theory and Applications 54, 2422–2437 (2007)
2. Garnett, E., Liang, W., Yang, P.: Growth and Electrical Characteristics of Platinum-Nanoparticle-Catalyzed Silicon Nanowires. Advanced Materials 19, 2946–2950 (2007)
3. Dehon, A.: Nanowire-based Programmable Architectures. J. Emerg. Technol. Comput. Syst. 1, 109–162 (2005)
4. Lu, W., Lieber, C.M.: Semiconductor Nanowires. Journal of Physics D: Applied Physics 39, R387–R406 (2006)
5. Wu, Y., Xiang, J., Chen, Y., Lu, W., Lieber, C.M.: Single-crystal Metallic Nanowires and Metal/Semiconductor Nanowire Heterostructures. Nature 430, 61–65 (2004)

Hybrid DNA and Enzyme Based Computing for Address Encoding, Link Switching and Error Correction in Molecular Communication

Frank Walsh[2], Sasitharan Balasubramaniam[1,2], Dmitri Botvich[1,2], Tatsuya Suda[3], Tadashi Nakano[3], Stephen F. Bush[4], and Mícheál Ó Foghlú[1,2]

[1] Telecommunication Software and Systems Group
{sasib,dbotvich,mofoghlu}@tssg.org
[2] Waterford Institute of Technology
Carriganore Campus, Ireland
{fwwalsh}@wit.ie
[3] Information and Computer Science
University of California, Irvine, CA, USA
{suda,tnakano}@ics.uci.edu
[4] Computing and Decision Sciences
GE Global Research
Niskayuna, NY, USA
bushsf@research.ge.com

Abstract. This paper proposes a biological cell-based communication protocol to enable communication between biological nanodevices. Inspired by existing communication network protocols, our solution combines two molecular computing techniques (DNA and enzyme computing), to design a protocol stack for molecular communication networks. Based on computational requirements of each layer of the stack, our solution specifies biomolecule address encoding/ decoding, error correction and link switching mechanisms for molecular communication networks.

Keywords: Molecular communication, molecular computing, communication protocols.

1 Introduction

In common with networked computing devices, biological cells have the ability to transmit, receive and process information through signaling networks and signal transduction mechanisms that interact in a complex biochemical system [6][7]. Just as modular silicon components are used to compose digital electronic circuits, the mechanisms that underpin biological systems are now being investigated to create a library of molecular components that can be used to engineer biological based nano (*bio-nano*) scale systems. One good example is Molecular Computing [12], which manipulates biomolecules to engineer biochemical based computing systems. By combining Molecular Computing and Molecular Communication [1], a new research

M. Cheng (Ed.): NanoNet 2008, LNICST 3, pp. 28–38, 2009.

domain that investigates bio-nano communication, the necessary computing mechanisms can be provided to create communication protocols for bio-nano devices (in the rest of this document, this will be referred to as *nanodevice*). Just as data communication protocols resulted in the rapid growth and ubiquity of networked computing devices and applications, the development of communication protocols for nano-based networks will stimulate groundbreaking future applications of bio-nano devices. The potential applications of these combined technologies are vast, particularly in the medical field where nano-scale devices can perform surgical procedures [14] or ensure accurate drug delivery to specific parts of organs and tissues.

Biological cells contain various components that can play vital roles in networked communication. These include, for example network interfaces (receptors, gap junctions), computing processes (regulatory networks, enzymatic signaling pathways) and memory capabilities (nucleic acids). In this paper, we propose a cell-based communication platform that uses these functional complexities to create protocols necessary for molecular communication networks. Our proposed hybrid solution includes DNA as well as enzyme based computing, where each contributes to specific protocol functions. We will describe how we will re-use protocols from communication networks, and transfer their mechanisms to a cell-based environment. In particular, we will show how our molecular communication protocol stack can support addressing, error correction, and link switching.

The paper is constructed as follows: Section 2 reviews the background of molecular communication and computing. Section 3 investigates protocols for data communication and how we reuse some of these concepts for our proposed protocols for molecular communication. Section 4 presents a simple connectionless communication solution using biological cells as a communication platform for address encoding, error correction, and link switching. Finally, section 5 presents conclusions and future work.

2 Background

2.1 Molecular Communication

Molecular Communication uses encoded molecules as information carriers to engineer biochemical-based communication systems. In [9], Moritani et al define a Molecular Communication Interface that uses vesicles embedded with gap junction proteins to transport message-encoded molecules. The vesicles that embed the information molecules (e.g. this could be represented as metabolites, or small nucleotides) will then be used as signal carriers between the sender and receiver nano-devices. Another form of molecular communication exploits the current calcium signalling that occurs between cells. For example, in [10] Nakano et al showed that distant nanodevices can communicate by encoding information through the frequency and amplitude of inter-cellular calcium waves.

2.2 Molecular Computing

This section will describe two common molecular computing techniques which include DNA and enzyme based computing. A summary and the characteristic differences between the two types of computation are also described.

2.2.1 DNA Based Computing

DNA is the universal "information molecule" and has a number of advantages in the computing world, such as encoding information as sequence of biochemical symbols as well as using these symbols to perform computing operations. In [3], Benenson et al present a programmable autonomous finite state automaton consisting entirely of biomolecules. The authors' design consists of a long DNA input molecule that is processed repeatedly by a restriction enzyme, and short DNA "rule" molecules that control the operation of the restriction enzyme. This concept forms the basis for a nanoscale computing machine that diagnoses disease and releases treatment molecules based on several disease-indicating inputs [11]. In [17], Liu et al extended the molecular automaton presented in [11] to design a "DNA-based Killer Automaton" that can release cytotoxic molecules which propagate to neighboring cells via gap junction channels.

2.2.2 Enzyme Based Computing

Markevich et al [4] created a bistable switch using a cell-based Kinase-Phophatase signaling cascade (MAPK) that is highly conserved in eukaryotic cells. In doing so, the author demonstrates the use of ultra-sensitive cell-based enzyme signaling pathways to perform digital logic computation. Similarly, in [5] Stetter et al uses the bistable nature of biochemical enzymatic reactions to create a reusable, "easy to engineer" architecture that forms the basis of several Boolean logic functions such as AND, and OR gates. This small enzyme-based circuit can act as a sub-component in composing more complex functions.

There are a number of differences between the two types of cell-based computing, where each has certain disadvantages and advantages with respect to computing for communication protocols. Firstly, the computational complexity and speed associated with DNA computing is, as yet, not attainable using enzyme based computing [16]. Also, the parameter characterization effort required to achieve enzyme computing increases dramatically relative to circuit complexity [13]. This makes enzyme computing more suitable for relatively simpler circuits that require short computation time. On the other hand, DNA-based computing can support larger computing requirements. The other difference between enzyme and DNA computing is that enzymatic reactions are intrinsic in cytosolic cell signaling pathways [7]. Therefore, this allows closer interaction with cell membrane components such as receptors and gap junctions. This makes it particularly suitable to simpler, responsive computing involving extra-cellular input and output.

3 Defining Protocols for Molecular Communication

In this section we will first describe the core characteristics of communication network protocols, and how these protocols will be re-used to support nanodevices.

3.1 Communication Network Protocols

Communication networks consist of protocols that exhibit the following properties; access mechanisms to physical communication interfaces, encoding and addressing mechanisms, error detection/correction techniques, and routing of packets between connected nodes. Physical interface controllers provide connection to physical transmission media and include mechanisms such as modulation and channel coding. The link layer functions manage access to the underlying physical layer, while flow control and acknowledgment mechanisms are usually implemented in higher layer protocols such as TCP. Communication can be connectionless or connection-oriented, where connectionless communication have lower data overhead, and are suitable for energy efficient networks such as wireless sensor networks. Another common protocol used in communication network is error correction, where techniques such as Forward Error Correction (FEC) can ensure that end devices can recover from any data corruption incurred during transmission. One approach is through inclusion of redundancy in channel encoding process.

3.2 Protocols for Molecular Communication

As described earlier, our intention is to be able to re-use protocols from conventional communication networks for molecular communication. Fig. 2 illustrates the components of our protocol stack and the protocols for different operations of the nanodevice (e.g. Transmitting node, Receiving Node, Intermediate Routing Node). Our approach is based on interconnection of loose protocol components, where each component is performed by a specific molecular computing technique. The reason that we have not embed all components into a generic protocol stack, is to prevent unnecessary increase in computational complexity. Although, the components of each layer is mapped from conventional protocols used in communication networks, the layers of our protocol stack is re-organised to suit a number of characteristics found in molecular communication. For example, propagation of information in molecular communication is typically characterized as low speed and in an environment where the interconnecting links between nano devices use biological signaling mechanisms that are highly variable compared to standard communication networks [1][2]. These characteristics have repercussions for the design of protocols of molecular communication systems. Slow diffusion-based processes do not support the creation of high-speed switching functions common in conventional network devices that will require complex queuing mechanisms for packets. At the same time, due to high variability and harsh biological environment, the use of acknowledgements and retransmission of messages in the event of loss or corrupt packets may not lead to improved performance.

We anticipate two types of information transmissions used in molecular communications, which includes sensory data (data collected from nanodevices) and command data (instructions for nanodevices). Therefore, the transmission mechanism and protocols to be used will be highly dependent on the nature of the information. For example, for sensor data, we may use single paths with UDP-like transmission with no error correction. However, command information or high priority sensor data will be transmitted through redundant paths with error correction capabilities (e.g. FEC).

Since protocols can usually be defined through a Finite State Machine (FSM), we adopt a nano-logic circuit that is translated from a FSM to represent the different types of protocols. We then map the specific protocol to either DNA or enzyme based computing. Since each technique has its own characteristics, we apply and select the right

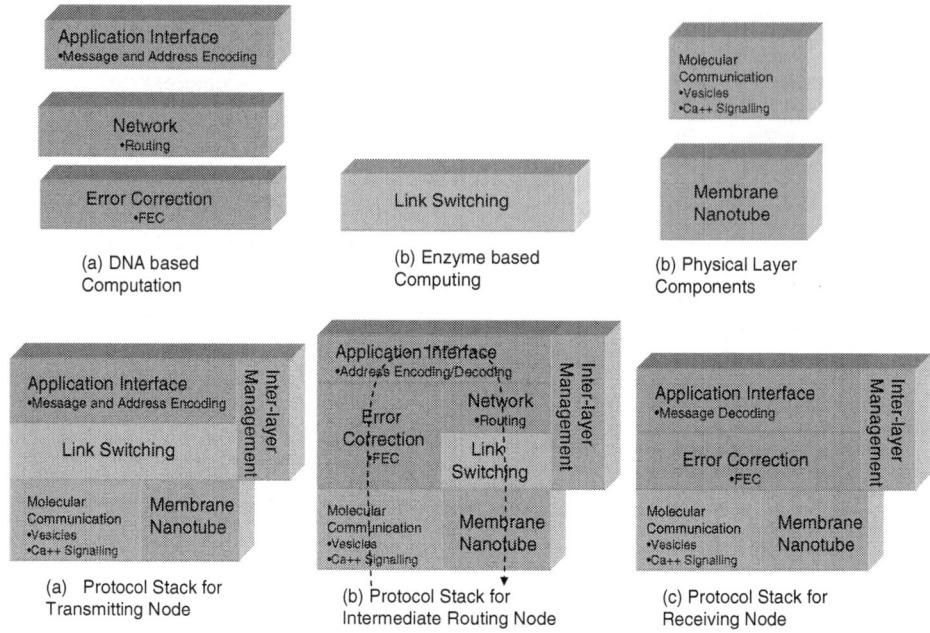

Fig. 1. Molecular Communication protocol stack

techniques based on two factors which includes, (i) the sequence of operation for the protocol, and (ii) complexity of computation required for the protocol. The DNA based computing is used for *Application Interface*, *Network*, and *Error Correction* layers, while the enzyme based computing is used for the *Link Switching* layer. The Application Interface, Network, and Error Correction layers will require higher complexity computation and is usually not required to be time sensitive. Such computations will include FEC, addressing, and information encoding/decoding. Enzyme based computing, due to its limited time requirement, is most suitable in performing small size logic circuit with high-speed computation. Therefore, this is most ideal for switching of information biomolecules between the links. The underlying physical layer can be based on solutions by [1] [10] for molecular communication, where the molecular communication can be guided through membrane nanotubes [19]. We select membrane nanotubes as a physical layer communication mechanism between cells, essentially providing the guided channels interconnecting each node in the bio-nano network. Unlike intercellular communication mechanisms that broadcast chemical signals to all neighboring cells via intercellular space, these nano-tubular structures can create a network of communication links between distinct cells that can support intercellular transfer of cytosolic molecules, vesicles and organelles. A notable work is by Önfelt et al who demonstrated a membrane network that transports tagged vesicles from cell to cell [19]. Therefore, the membrane nanotubes could be used in conjunction with a suitable molecular communication mechanism such as [9] that uses vesicles to transport message molecules or [10] to guide modulated calcium "waves" from sending cell to receiving cell.

In between the two layers will be the *Inter-layer protocol management*, which will coordinate the different computation of each layer of the protocols and the location where this will happen in the cell. Fig. 2 illustrates our solution that combines a subset of our proposed protocol to support transmission on a single link.

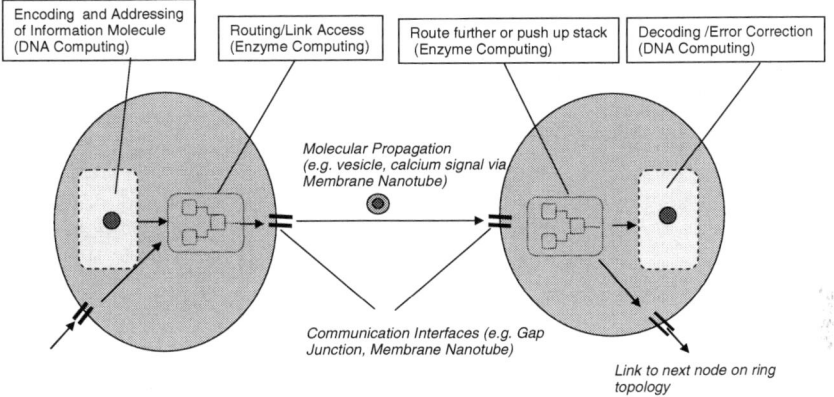

Fig. 2. Mechanism of transmission for single link molecular communication

The flow of operation between the layers is as follows. For the Transmitting Node (Fig. 1(a)), the Application layer Interface will perform the message encoding for the information biomolecules. The encoded biomolecule is then further encoded with the specific address of the intended destination using an address table. In our proposed protocol stack, we have left our application layer open, where the cells can interface to a physical device or we can have artificial cells with embedded functionalities (e.g. the cell also acts as the device). Once the encoding process is performed, the information biomolecule is ready for transmission and submitted to the Link Switching layer, which selects the correct gap junction for transmission. In the Intermediate Routing Nodes case (Fig. 1(b)), when the biomolecule is received by the cell, the error correction is first performed on the biomolecule. This is then followed by the address decoding and encoding process based on the routing table for the next node. Once this is performed, the link switching operation follows and transmits the biomolecule to the underlying link. Once the information biomolecule is received at the receiving device (Fig. 1(c)), the information biomolecule is once again passed through the Error Correction layer to perform any necessary error correction, which is then followed by the message decoding at the Application Interface layer.

4   Proposed Solution

In this section, we will describe the molecular computing operations for information encoding and addressing, link switching, as well as error correction.

4.1 Encoding and Addressing

Fig. 3(a) illustrates the encoding process. Similar to the model proposed by Liu et al in [17], our solution uses Benenson's and Shapiro's work in [3] to create a DNA-based automaton that produces a *single strand DNA (ssDNA)* message molecules for intercellular communication. Each ssDNA message is encoded as a unique sequence of nucleotide bases as demonstrated in [4]. For simplicity, only three addressable nano-device nodes are considered and each encoded ssDNA message is 'framed' to include addressing information.

Fig. 3(a) illustrates how nucleotide encoded messages are assembled in sequence of long input double stranded DNA message molecule with each message separated by a 'spacer' sequence. The upper leftmost "sticky end" represents the current state of the machine. During the address encoding process, the DNA message molecule is cut by a restriction enzyme, which releases the leftmost segment of the molecule. Thus the *<address, message>* pairing represented by the current state of the encoding automaton is released as an ssDNA segment through the restriction process. Fig. 3(b) illustrates how each address state and transition corresponds to actual encoded message molecule. Each state transition is enacted by a corresponding DNA "rule" molecule and enzyme complex that cleaves the corresponding nucleotide sequences. A key characteristic of address encoding is the precise cleaving of input message molecule that encodes or "frames" the message.

Fig. 4 illustrates a rule execution transition from Address 2 to Address 3. Each rule molecule has a recognition site to which a restriction enzyme can bind. As described earlier, the number of nucleotide bases between the restriction enzyme and the sticky end of the rule molecule determines the precise locations of the message molecule

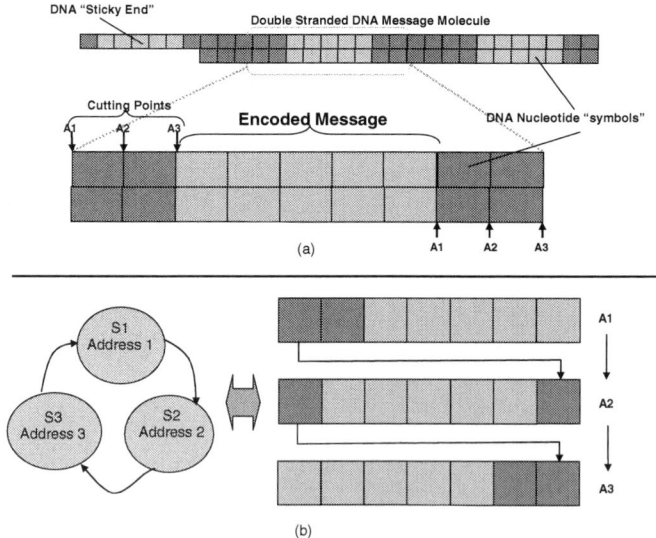

Fig. 3. (a) Double Stranded DNA message molecule indicating restriction cut points for address encoding, (b) State representation of address encoding transitions

cleave. In this example, the restriction enzyme complex combines with the message molecule and cuts at fourteen nucleotides on the top and twenty-one nucleotides at the bottom. The resulting new sticky end reveals the next state of the automaton. More importantly, the segment that is cut away separates into two ssDNA molecules. The lower ssDNA molecule indicated in Fig. 4 is the encoded message molecule with its rightmost end complementary to the new sticky end of the DNA message molecule.

Similar to techniques used in [17] and [11], the nanodevice can control computation by releasing molecules (e.g. mRNA) that selectively activate DNA *"rule"* molecules. The results of the computation can provide input to other parallel computational functions, which was proposed in [3]. In our solution, the cleaved ssDNA message molecules are released into the cytosol and provide the input to the molecular interface control function of the network layer. Theoretically, this mechanism can be extended to encode a multitude of unique address locations and any number of messages during computation.

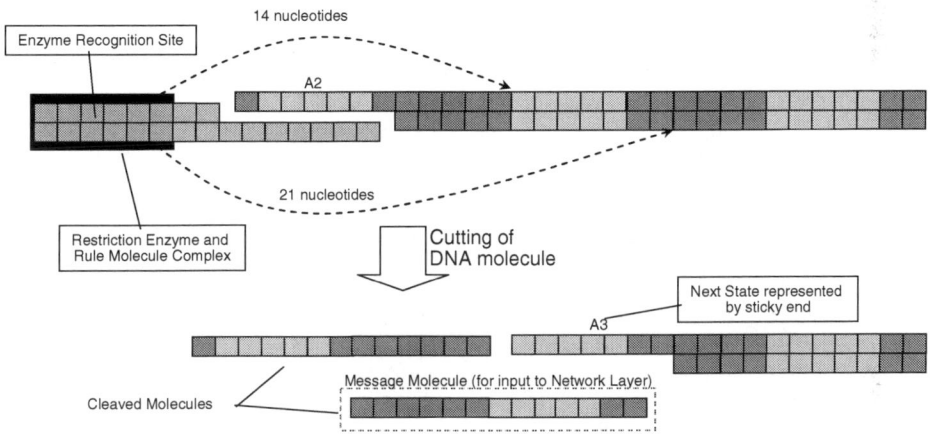

Fig. 4. Mechanism of State Transition from Address 2 to Address 3 using Benenson's Molecular Automata [3]

4.2 Molecular Interface Control

As described earlier, the operation of our molecular communication is through a membrane nanotube network. Fig. 5 illustrates a cell with two distinct molecular communication interfaces (e.g. distinct gap junctions). Each addressable location is switched through the corresponding interface according to the addressing state diagram shown in Fig. 5 (b). For communication involving the transfer of message molecules through gap junctions, our solution is based on results in [18] which demonstrate the diffusion of synthetic oligonucleotides through gap junction channels. In our case, instead of oligonucleotides, we diffuse our encoded ssDNA from the previous section.

In this study, interface selection is achieved using the "real world" implementation of the logical recurrent architecture as described by Stetter et al in [5]. The switching circuit releases/alters a corresponding chemical signal that "switches" the ssDNA to

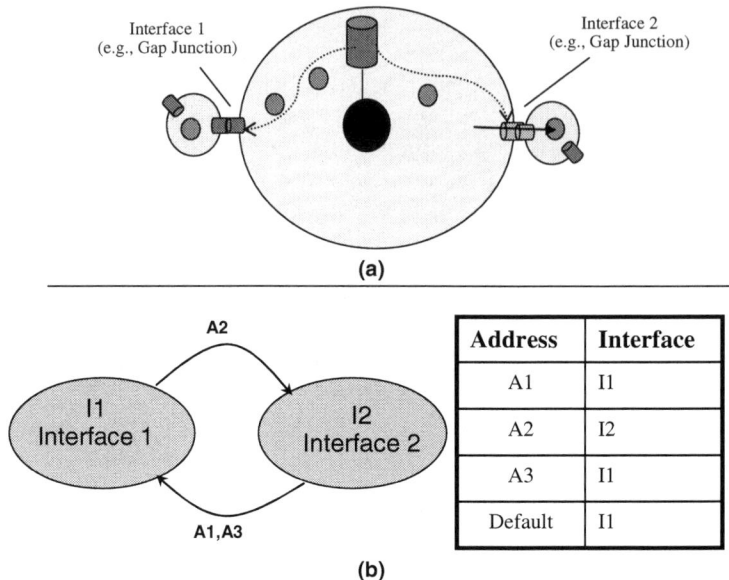

Fig. 5. (a) Schematic diagram of cell with two distinct molecular communication interfaces, (b) Address/Interface state diagram and switching table

the correct interface. In the case of gap junction interfaces, the output of the enzyme-based circuit will control the permeability of gap junction channels. Gap junction permeability is affected by the connexin phosphorylation [10] via specific concentration of phosphorylation reagents.

Thus Stetter's circuit can be used to effectively switch on and off each molecular communication interface by controlling the degree of phosphorylation of gap junction connexins. This in turn will allow the ssDNA to be pushed through only a single link (or multiple links if multicasting is used). Using this technique, several communication links can be controlled simultaneously via compartmentalized enzymatic functions [8]. The Inter-layer protocol we will be responsible for triggering the enzymatic computation, once the operation from the Application Interface layer is complete (the operation of this mechanism is subject to future work).

4.3 DNA Decoding and Forward Error Correction

As already stated, prioritized messages require error detection and correction. Invariably, errors will occur in the encoding and transmission process of ssDNA molecules due to the imprecise nature of the associated complex biochemical reactions [15]. By including redundancy in the encoding process, error correction mechanisms can be incorporated into the decoding process. Our solution combines the nucleotide redundancy concept presented in [16] with DNA automata design in [11] to create an autonomous error correction mechanism. In our proposed technique, each ssDNA molecule is composed of several repeated, identical nucleotide sequences.

In [11] Benenson uses "protector strands" to control the operation of an enzyme based state machine by separating the constituent DNA strands of message molecules (see Fig. 6). In our solution, the protector strands are designed to have a strong affinity for a specific received ssDNA. The ssDNA molecules cause the corresponding protector strand to separate from the transition strand and hybridize with the message molecule allowing the formation, and thus activation, of a double stranded transition molecule (a similar mechanism to the encoding process). The resulting transition molecule and restriction enzyme complex cleaves the corresponding decoding DNA molecule and releases the decoded DNA molecule (in Fig. 6, this is represented as the end DNA hairpin) with no errors. Our assumption of this approach is mainly for finite instruction messages, where our end device will contain as many Decoding DNA molecule as the number of possible instructions. Hybridization can also occur even though both the protector strand and the received ssDNA molecule are not exactly complementary.

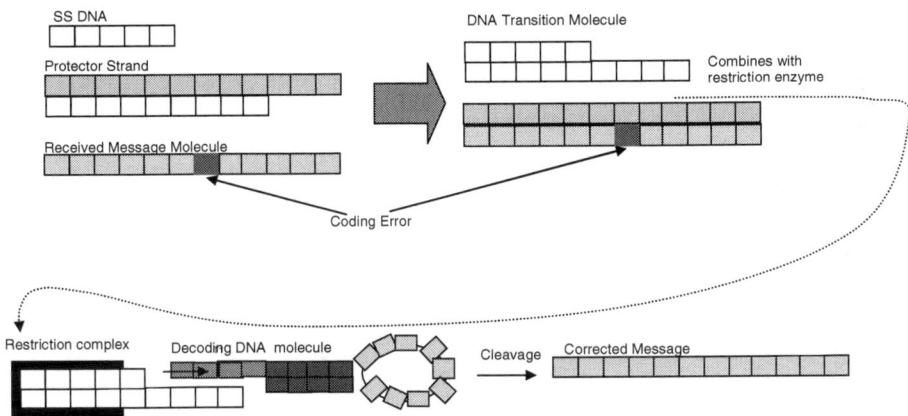

Fig. 6. Forward Error Correction Mechanism

5   Conclusion and Future Work

Inspired by protocols for communication networks, we have presented a molecular communication protocol stack that successfully combines molecular computing and molecular communication techniques. We describe how the core characteristics of communication network protocols are re-used to design bio-nano device communication protocols. Our proposed protocol stack presents the address encoding/decoding, link switching, and error correction functions that are developed using molecular computing techniques. The solution demonstrates the necessity of matching the characteristics of each molecular computing technique to the computational requirements of each layer of the proposed protocol stack. Our future work will investigate the feasibility of our design initially through simulation of chemical circuits for molecule encoding/decoding, link switching and error correction.

Acknowledgements. This work has received support from the Higher Education Authority in Ireland under the PRTLI Cycle 4 programme, in a project Serving Society: Management of Future Communications Networks and Services.

References

1. Hiyama, S., Moritani, Y., Suda, T., Egashera, R., Enomoto, A., Moore, M., Nakano, T.: Mole-cular Communication. In: Proc. NSTI Nanotech Conference and Trade Show, Anaheim, vol. 3, pp. 391–394 (May 2005)
2. Alfano, G., Morandi, D.: On Information Transmission Among Nanomachines. In: 1st International Conference on Nano-Networks (Nano-Net), Lousanne, Switzerland (2006)
3. Benenson, Y., Shapiro, E.: Molecular Computing Machines. In: Dekker Encyclopaedia of Nanoscience and Nanotechnology, pp. 2043–2055. Marcel Dekker, New York (2004)
4. Markevich, N., Hoek, J., Kholodenko, B.: Signalling switches and bistability arising from multisite phosphorylation in protein kinase cascades. J. Cell Biology 164(3), 353 (2004), http://www.jcb.org/cgi/reprint/164/3/353
5. Stetter, M., Schurmann, B., Hofstetter, M.: Logical Nano-Computation in Enzyme Reaction Networks. In: Proceedings of BIONETICS 2006, Cavalese, Italy (2006)
6. Sauro, H., Kholodenko, B.: Quantitative analysis of signalling networks. Prog, in Biophysics & Mol. Biology 86, 5–43 (2004)
7. Cooper, G.: The cell, a molecular approach, pp. 545–549, ISBN 0-87893-119-8
8. Komarova, N., Zou, X., Nie, Q., Bardwell, L.: A theoretical framework for specificity in cell signalling. Mol. Syst. Biol. (October 2005)
9. Moritani, Y., Hyama, S., Suda, T.: A Communication Interface Using Vesicles Embedded with Channel Forming Proteins in Molecular Communication. In: Proceedings of BIONETICS 2007, Budapest, Hungary (2007)
10. Nakano, T., Suda, T., Koujin, T., Haraguchi, T., Hiraoka, Y.: Molecular Communication through Gap Junction Channels: System Design, Experiments and Modeling. In: Proceedings of 2nd International Conference on Bio-inspired models of Network, Information, and Computing Systems, Budapest, Hungary (2007)
11. Benenson, Y., Gil, B., Ben-Dor, U., Adar, R., Shapiro, E.: An Autonomous Molecular Computer for Logical Control of Gene Expression. Nature 429, 423–429 (2004)
12. De Silva, A.P., Uchiyama, S.: Molecular Logic and Computing. Nature Nanotechnology 2 (July 2007)
13. Niazof, T., Baron, R., Katz, E., Ioubashevski, O., Willner, I.: Concatenated logic gates using four coupled biocatalysts operating in series. In: PNAS, Niazov, November 14 (2006)
14. Cavalcanti, R., Freitas Jr., A.: Nanorobotics Control Design: A Collective behaviour approach for medicine. IEEE Transaction on NanoBioscience 4(2) (June 2005)
15. Abelson, H., Allen, D., Coore, D., Hanson, C., Homsy, G., Knight, T.F., Nagpal, R., Rauch, E., Sussman, G., Weiss, R.: Amorphous Computing. Communications of the ACM 43(5) (May 2001)
16. Fedichkin, L., Katz, E., Privman, V.: Error Correction and Digitalization Concepts in Biochemical Computing. J. Comput. Theor. Nanosci. 5, 36–43 (2008)
17. Liu, S., Gaudiot, J.: DNA-based Killer Automaton: the Innovative Nanomedicine. In: Proc. NSTI Nanotech Conference and Trade Show, Boston, Massachusetts (May 2006)
18. Valiunas, V., Polosina, Y.Y., Miller, H., Potapova, I.A., Valiuniene, L., Doronin, S., Mathias, R.T., Robinson, R.B., Rosen, M.R., Cohen, I.S., Brink, P.R.: Connexin-specific cell-to-cell transfer of short interfering RNA by gap junctions. J. Physiol. 568(2), 459–468 (2005)
19. Önfelt, B., Nedvetzki, S., Benninger, R.K.P., Purbhoo, M.A., Sowinski, S., Hume, A.N., Seabra, M.C., Neil, M.A.A., French, P.M.W., Davis, D.M.: Structurally Distinct Membrane Nanotubes between Human Macrophages Support Long-Distance Vesicular Traffic or Surfing of Bacteria. FEBS Lett. 581(11), 2194–2201 (2007)

Hitting Time Analysis for Stochastic Communication

Paul Bogdan and Radu Marculescu

Carnegie Mellon University, Pittsburgh, PA, USA

Abstract. This paper investigates the benefits of a recently proposed communication approach, namely *on-chip stochastic communication*, and proposes an analytical model for computing its mean hitting time. Towards this end, we model the stochastic communication as a branching process taking place on a finite mesh and estimate the mean number of communication rounds.

Keywords: Network-on-Chip, reliable communication, hitting time.

1 Introduction

Shrinking geometries, scaling down supply voltages, and increasing clock frequencies have a negative impact on System-on-Chip (SoC) reliability [8]. Thus, there is a great need for scalable and reliable communication protocols among the SoC components. The traditional acknowledgement/request protocols are not adequate in such an error prone environment. To mitigate the impact of unpredictable faults in the deep submicron domain, a biologically-inspired communication was proposed in [4]. In that paper, the authors quantify the stochastic communication node coverage, but do not evaluate concrete performance metrics (*e.g.*, mean hitting time). However, evaluating the hitting time for a source-destination pair of nodes is important as it can serve as an input parameter for task mapping and scheduling problems in SoCs.

Starting from these ideas, this paper provides a theoretical framework for computing the mean hitting time between any two nodes in a mesh under stochastic communication protocol. We model the stochastic communication as a branching process in which each node that receives a copy of the disseminated packet, duplicates and sends it probabilistically to a subset of its neighboring nodes. In contrast to a *single* random walk, where a message is sent randomly only to a neighbor at any given time, the stochastic communication consists of *multiple* random walks that start at each node and provide a higher dissemination speed and robustness to the communication protocol. A major issue raised by the hitting time analysis for stochastic communication is that the diffusion process takes place on finite cyclic graphs. Also due to link or node failures, some of these random walks end prematurely. We model this behavior as an annihilation process. Thus, not only the topology, but also the protocol features consisting of branching and annihilation phases, make the hitting time analysis difficult. By estimating the hitting time, we gain insight into the consequences of link/node failures on node-to-node communication delay; this can help us adjust the forwarding probability such that the destination node receives at least one copy.

M. Cheng (Ed.): NanoNet 2008, LNICST 3, pp. 39–43, 2009.
© ICST Institute for Computer Sciences, Social Informatics and Telecommunications Engineering 2009

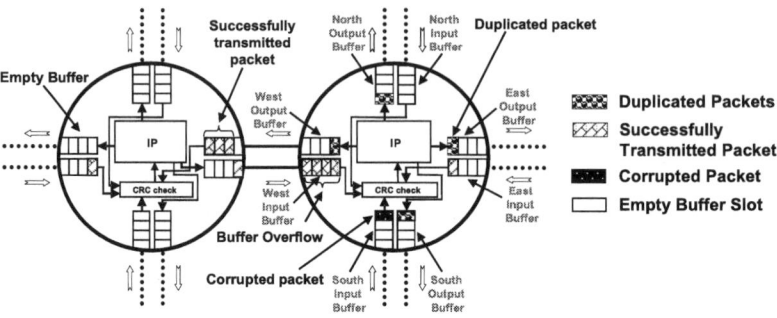

Fig. 1. Possible events between two neighboring nodes under stochastic communication.

The paper is organized as follows: Section 2 presents an overview of the hitting time theory and motivates the hitting time analysis for multiple random walks. Section 3 describes the Markov chain model associated with the stochastic communication and how the mean hitting time can be estimated. Section 4 presents our experimental results, while Section 5 outlines possible directions for future work.

2 Related Work

The hitting time concept for random walks was developed in connection with electrical networks [7][10] and attracted significant attention due to its potential applications such as the design of distributed computation [5][7][11], estimation of the complexity of distributed algorithms [6], search in peer-to-peer networks [12], estimation of Web size [3]. While the evolution of a random walk on graphs is extensively studied, only recently there has been an increased interest in the study of multiple, yet finite number of random walks [1][9]. However, many natural phenomena (*e.g.*, epidemics [2]) and human made processes cannot be modeled by imposing a single or finite number of random walks. Thus, similarly to epidemics, we model the stochastic communication as a *collection* of random walks which increase or decrease in cardinality based on the forwarding probability or the packet corruption probability.

3 Hitting Time Analysis

We consider a stochastic communication scenario in an $N \times N$ mesh network starting at node $(i,j)_S$. The protocol evolves as follows: If the packet is successfully received at any given node, it is first CRC checked for information integrity. If the packet is not already a duplicate, the current node copies and probabilistically sends it to a set of its neighbors. Two design cases can be considered: Either the packets are stored at each node and the transmission between two neighbors happens only when there is a free slot (*e.g.*, the West input buffer of right hand side node in Fig. 1 is full and no new packets can be stored - a buffer overflow is flagged), or the protocol allows the packets stored in the buffer to be overwritten by the incoming packets. For the sake of simplicity, we assume that packets can be overwritten and discuss only how the buffer overflow can be modeled. Besides these events, the packets can be lost due to node failures (*e.g.*, the incoming packet on the South input buffer of the right hand side node of Fig. 1 can be corrupted during the routing decision). The packet diffusion is a branching process, while the reverse process, in which packets are corrupted or over-written, is an *annihilating* process. The number of received copies at a given node (i,j) can be described via a Markovian process $\{O_{ij}(t)|1 \leq i,j \leq N, t \geq 0\}$ as follows:

a) Packet duplication: In a short time interval δt, a node (i,j) can duplicate a packet (see the right hand side node in Fig. 1) according to the next relation:

$$Pr\{O_{ij}(t+\delta t)=k+1|O_{ij}(t)=k\} = \lambda_{ij}k\delta t+O(\delta t) \qquad (1)$$

where λ_{ij} is the packet duplication rate for each node (i,j), k is the number of received copies, and $O(\delta t)$ is a negligible term. The duplication starts only if the node (i,j) received at least 1 copy $(k > 0)$, otherwise the probability is zero.

b) Packet successful transmission: A packet duplicated at node (i,j) can be suc-cessfully sent to its neighbor $(i,j-1)$ (thus increasing its O_{ij-1}) as follows:

$$Pr\{(O_{ij}, O_{ij-1})(t+\delta t)=(k-1, n+1)|(O_{ij}, O_{ij-1})(t)=(k, n)\} = \alpha_{ij}^{W}k\left(1 - \frac{n}{B_{ij-1}}\right)\delta t+O(\delta t) \qquad (2)$$

where α_{ij}^{W} is the link successful transmission rate from node (i,j) to its West neigh-bor $(i,j-1)$ (*i.e.*, the router at (i,j) sends a copy to $(i,j-1)$ node and the packet is success-fully received). This probability is strictly positive only if the sending node (i,j) has received at least one copy. The excessive packet duplication can cause buffer over-flow situations, which can be captured by inserting the $(1 - O_{ij-1}/B_{ij-1})$ term, B_{ij-1} being the buffer size at the $(i,j-1)$ node. If the buffer at node O_{ij-1} is empty, then this term has no effect on the transition probability. If the number of received copies O_{ij-1} increases, then this probability decreases. Similarly, we can describe the transmission events from (i,j) to the North (O_{i-1j}), East (O_{ij+1}), and South (O_{i+1j}) neighbors.

c) Packet corruption while routing: The probability that a node corrupts the received packet during the routing decision, causing it to be discarded, is:

$$Pr\{O_{ij}(t+\delta t)=k-1|O_{ij}(t)=k\} = \mu_{ij}k\delta t+O(\delta t) \qquad (3)$$

where μ_{ij} is the packet corruption rate at node level. This transition is activated with rate μ_{ij} only if the node (i,j) already received a positive number of packets $(k > 0)$. This accounts for potential errors in computation at node-level.

The packet diffusion over the entire network is seen as a *collection* of Markov pro-cesses, where the occurrence probability of any transition in an interval $(t,t+\delta t]$ relies on the number of packets received at time t in each node. The nodes evolution can be described via a master equation of the multivariate probability distribution:

$$P(o_{11}..., o_{ij}..., o_{NN};t)=Pr\{O_{11}(t)=o_{11}, ...O_{NN}(t)=o_{NN}|O_{11}(0)=m_{11},...O_{NN}(0)=m_{NN}\} \qquad (4)$$

which shows that the stochastic process $O_{ij}(t)$ received o_{ij} packets $(O_{ij}(t) = o_{ij})$. The evolution of probability function (*i.e.*, Eq. 4) is given by the following equation:

$$\frac{dP(...o_{ij}...;t)}{dt} = \sum_{i,j=1}^{N} [\lambda_{ij}(o_{ij}-1)P(...o_{ij}-1,...;t)+\mu_{ij}(o_{ij}+1)P(...o_{ij}+1...;t)$$

$$+\alpha_{ij}^{N}(o_{ij}+1)\left(1-\frac{o_{i-1j}-1}{B}\right)P(...o_{i-1,j}-1,o_{ij}+1;t)+\alpha_{ij}^{E}(o_{ij}+1)\left(1-\frac{o_{ij+1}-1}{B}\right)P(...o_{ij}+1,o_{ij+1}-1;t)$$

$$+\alpha_{ij}^{E}(o_{ij}+1)\left(1-\frac{o_{ij+1}-1}{B}\right)P(...o_{ij}+1,o_{ij+1}-1;t)+\alpha_{ij}^{W}(o_{ij}+1)\left(1-\frac{o_{ij-1}-1}{B}\right)P(...o_{ij}+1,o_{ij-1}-1;t) \quad (5)$$

$$+\alpha_{ij}^{S}(o_{ij}+1)\left(1-\frac{o_{i+1j}-1}{B}\right)P(...o_{ij}+1,o_{i+1,j}-1...;t)] -$$

$$-\sum_{i j=1}^{N}\left[\lambda_{ij}+\mu_{ij}+\alpha_{ij}^{N}\left(1-\frac{o_{i-1j}}{B}\right)+\alpha_{ij}^{E}\left(1-\frac{o_{ij+1}}{B}\right)+\alpha_{ij}^{W}\left(1-\frac{o_{ij-1}}{B}\right)+\alpha_{ij}^{S}\left(1-\frac{o_{i+1j}}{B}\right)\right]o_{ij}P(...o_{ij}...;t)$$

with an initial condition $P(o_{11}=0..., o_{(ij)}=1,_s...o_{NN}=0;t=0) = 1$.

Traditional methods of solving Eq. 5 (*e.g.*, the moment generating function, the transition matrix approach) do not offer a scalable solution. We approximate the solution of Eq. 5 via nonhomogeneous Poisson distribution [13] and express the mean hitting time between source $(i,j)_S$ and destination $(i,j)_D$ as follows:

$$\langle t \rangle = \int_0^\infty t \sum_{o_{(ij)_D}=0}^{1} P(o_{(ij)_D}, t; o_{(ij)_S}= 1, t = 0)dt \quad (6)$$

where the right hand side accounts for the mean time during which the destination node did not receive any packet and the time needed to receive exactly one packet.

4 Experimental Results

Next, we evaluate the proposed model by considering a stochastic communication scenario between node $(1,1)_S$ and node $(10,10)_D$ on a 10×10 mesh NoC. First, we estimate the probability for node $(10,10)$ to receive exactly 0, 1, 2, 3 and 4 packets (see

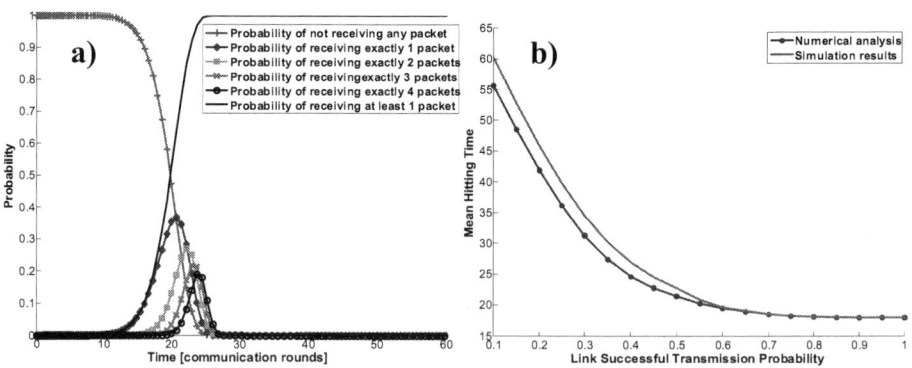

Fig. 2. a) Time-dependent probabilities for destination (10,10) to receive 0, 1, 2, 3 and 4 packets from source (1,1). The probability of receiving at least 1 copy is almost 1 after 25 communication rounds. b) Comparison between analytical and simulation results of the mean hitting time between (1,1) and (N,N) nodes on a 10×10 mesh.

Fig. 2.a. We also report the time-dependent probability of receiving at least 1 packet. For this experiment, we use a 0.15 packet injection rate, and a 0.8 link successful transmission probability for all directions. The probability of receiving k or more copies at destination can be used for designing various voting strategies to improve the error correcting methods (*e.g.*, the destination can recover the packet out k received copies). Fig. 2.b shows the mean hitting time between nodes $(1,1)$ and (N,N) obtained via Eq. 6 and simulation in a 10×10 mesh, a 0.15 packet duplication rate, and zero probability of overflow. The mean hitting time results were obtained by averaging over 500000 runs. We note that the numerical analysis becomes more inaccurate compared to the simulation as the link successful transmission probability decreases.

5   Conclusions

This paper presented a framework for computing the mean hitting time of a branching process running on a finite graph and discussed its application to stochastic communication. As future work, we plan to model the traffic burstiness and the effects of finite buffers so that the analysis provides accurate estimates of the node-to-node latency.

Acknowledgements. Authors acknowledge the support of the Gigascale Systems Research Focus Center, one of the five research centers funded under the Focus Center Research Program, a Semiconductor Research Corporation program.

References

1. Alon, et al.: Many Random Walks Are Faster Than One. In: 20th ACM SPAA, Germany (2008)
2. Bailey, N.T.J.: The Mathematical Theory of Infectious Diseases (1975)
3. Bar-Yossef, Z., Gurevich, M.: Random sampling from a search engine's index. In: WWW (2006)
4. Bogdan, P., Dumitras, T., Marculescu, R.: Stochastic Communication: A New Paradigm for Fault-Tolerant Networks-on-Chip. In: Hindawi VLSI-Design (Feburary 2007)
5. Broder, A., Karlin, A.R.: Bounds on Cover Time. J. of Theor. Prob. 2(1) (1989)
6. Bui, A., Sohier, D.: On Time Analysis of Random Walk Based Token Circulation Algorithms. In: Ramos, F.F., Larios Rosillo, V., Unger, H. (eds.) ISSADS 2005. LNCS, vol. 3563, pp. 63–71. Springer, Heidelberg (2005)
7. Chandra, et al.: The Electrical Resistance of a Graph Captures its Commute and Cover Times. In: ACM Symp. on Theor. of Computing (1989)
8. Constantinescu, C.: Trends and Challenges in VLSI Circuit Reliability. IEEE Micro (2003)
9. Cooper, C., Frieze, A., Radzik, T.: Multiple Random Walks in Random Regular Graphs. CMU Technical Report (May 2008)
10. Doyle, P.G., Snell, L.J.: Random Walks and Electrical Networks, M.A.A (1984)
11. Eugster, et al.: Epidemic information dissemination in distributed systems. Computer (2004)
12. Gkantsidis, A., et al.: Random walks in peer-to-peer networks. In: INFOCOM (2004)
13. Solari, H.G., Natiello, M.A.: Poisson approximation to density dependent stochastic processes. A numerical implementation and test. Växjö Univ. Press., Math. Modelling (2003)

FPAA Based on Integration of CMOS and Nanojunction Devices for Neuromorphic Applications

Ming Liu[1], Hua Yu[2], and Wei Wang[2]

[1] Institute of Microelectronics, Chinese Academy of Science, Beijing, China
[2] College of Nanoscale Science and Engineering,
State University of New York at Albany, 255 Fuller Rd., Albany, NY, USA
{stanachutiwat,wwang}@uamail.albany.edu

Abstract. In this paper, a novel field programmable analog arrays (FPAA) architecture, namely, NueroFPAA, is introduced to utilize nanodevices to build a programmable neuromorphic system. By using nanodevices as programmable components, the proposed FPAA can achieve high-density and low-power operations for neuromorphic applications. The routing and function blocks of the FPAA are specifically designed so that this proposed architecture can support large-scale neuromorphic design as well as various analog circuitries.

Keywords: Field programmable analog arrays (FPAA), Nanojunction devices, Operational amplifier (Op-amp).

1 Introduction

Field-programmable analogue array (FPAA) is a reconfigurable platform to build analog circuits and can dramatically reduce time-to-market in analog circuit development [1-4]. One promising application of FPAA is to build a neuromorphic system that can mimic human brain. Such intelligent computing systems can carry out analog computation with extremely low power consumption, which can outperform the traditional digital computers for various applications such as pattern recognition and classification [4-8]. Even though FPAA is an efficient reconfigurable platform to establish electronic neuromorphic machine, the device density and complexity of the current FPAA needs to be significantly increased. The typical neuromorphic system scalable to biological levels will require a density of 1010 nodes/cm2 and a complexity of 1014 components. In order to achieve such a high density and large complexity, new devices and architectures of FPAA need to be developed.

In this paper, we introduce a new FPAA structure, namely, nueroFPAA, by utilizing CMOS devices as logics and emerging memory devices as programming elements. In particular, we use resistive junction to build the programmable elements in both routing channel and function block of a FPAA. The nanojunction device is a metal-insulator-metal nanojunction with hysteretic resistance characteristics. The routing channel based on these junctions can work as the synapses reaching a density of 1010/cm2 [5-8]. The function block based on a combination of CMOS and junction devices will be a reconfigurable component as neurons. Therefore, the proposed

M. Cheng (Ed.): NanoNet 2008, LNICST 3, pp. 44–48, 2009.
© ICST Institute for Computer Sciences, Social Informatics and Telecommunications Engineering 2009

system not only can support the complex analog applications such as filters, differentiators, but also can provide an efficient platform to build neuromorphic system that mimic human brain.

2 Proposed Neuro-FPAA Architecture

In this section, we will first review the existing FPAA structure [1] and then develop a new neuroFPAA architecture to utilize the nanojunction devices with CMOS devices for neuromorphic applications.

2.1 FPAA Basics

FPAA mainly consists of: (1) routing channel that connects the function blocks. (2) Functional blocks or configurable analog blocks that implement circuit functions. Both routing channel and function block have the reconfigurable capabilities.

The routing channel is based on the global and local crossbars. At each crosspoint of the crossbar, a floating gate is required to provide reconfigurable capabilities. The CAB consists of Op-amp, detector, SOS, matrix multiplier, peak detector and capacitors. The configurable capabilities of CAB come from the local crossbar that connects the CAB.

This existing FPAA architecture is sufficient for analog circuit development. However, for the neuromorphic application, the size of the crossbar is increasing. The size of crossbar will be considerably large due to the use of floating gate transistors. This motivates us to use nanojunction to replace floating gate to achieve a high-density FPAA.

2.2 Proposed Neuro-FPAA

As shown in Fig. 1, the proposed FPAA architecture contains routing crossbars based on nanojunctions and CABs incorporating nanojunctions inside to obtain CAB programmable capabilities. The arrangement of routing and CAB in Fig. 1 is specifically suitable for neuromorphic applications.

Nanojunction-Based Routing for Synaptic Operations: Fig. 1 shows the feed forward case such that each CAB will reach two following CABs through the routing crossbar. Each CAB consists of 10 neurons and the size of the routing crossbar is 38*19.

By utilizing the crossnet concept, the neural network applications will require the presynaptic neuron to reach the postsynaptic neuron using two different paths. For example, the same output signals of a neuron in CAB-I will reach a neuron of CAB-II in two paths. Using two other paths, CAB-I will also reach CAB-III. Therefore, each neuron can reach 20 neurons, requiring 38 paths.

If the recurrent neural network is implemented, each forward path will have a corresponding backward path that is in parallel to the forward path with a reverse direction. Thus, the crossbar will be duplicated for this purpose.

Nanojunction-Based CAB as Neuron: The CAB consists of 10 neurons. Each neuron is essentially an Op-amp with R, C supporting elements. The Op-amp and its internal

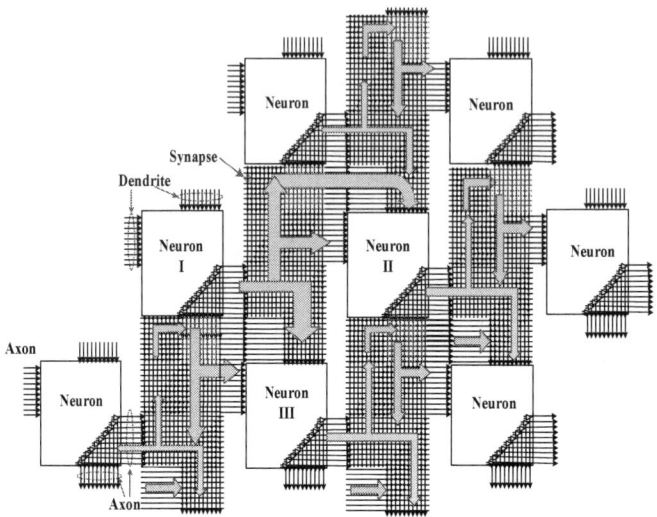

Fig. 1. Proposed neuroFPAA architecture

arrangement strongly affect the flexibility and functionality of the neuron. Thus, we propose a CAB with 10 Op-amps and several programmable capacitor arrays (PCAs), programmable resistor arrays (PRAs). The nanojunctions are also included in the PRAs and PCAs to provide programmable capabilities (see Fig. 1).

Note that the PCA and PRA are not only useful the neuron application but also can realize feedback loop, signal coupling, integration, differentiation and other analog signal processing functions. In this way, this proposed CAB structure can fulfill the requirement of general FPAA analog circuit applications as well as neuromorphic applications.

Improved Op-amp in CAB: The Op-amp is the core building block for a neuroFPAA. It should have adaptability and flexibility. In our design, applications of operational amplifiers include non-linear circuit's application and linear circuit's application. Non-linear applications include: neuron, logarithmic amplifier, and exponential amplifier. Linear applications include: voltage to current converters, current to voltage converters, summing amplifier inverter, noninverter, the integrator, and the differentiator.

We specifically improve the Op-amp design to achieve high performance both types of applications. It is a two-stage design consisting of a folded-cascade amplifier, source follower and compensation network. By using this improved design, the properties of Op-amp in terms of gain and phase margin, power supply rejection ratio and common-mode rejection ratio are significantly improved.

3   Operation Analysis and Performance Evaluation

The operation of the proposed FPAA as a neural network consists of two steps: training and operation. To statically train the routing crossbar, the dedicated programming circuitry (not shown in Fig. 1) is required to configure the synaptic junction a prior. This is generally referred to pre-computation training. The second method is called "in situ" training. Since the junction is a two terminal device, we can also modulate the current and voltage of neuron to program the junction dynamically.

After the training or configuration step is over, the neuroFPAA is changed to operation mode. The input signals are inserted into the network and allow the synapses and neurons to start operations.

During the operation, the system can have defect tolerance and low-power performance. The junction devices may have high defect rates. The nanowire connections can also be defective. However, since NN can function with these defective devices while maintaining minimal performance degradation, the neuromorphic system can provide high defect tolerance: in some cases, it can provide 99% fidelity with more than 80% fraction of bad devices [8].

The low-power properties of the system stem from the low working frequency. The system is generally working in 100HZ and KHz ranges [9], leading to low power consumption.

3.1 Performance Evaluation

The performance analysis is carried out based on the estimation of nanojunction performance, which will be compared with floating gate devices. The 38*19 crossbar will require 722 transistors (T) in the floating gate based FPAA structure. By using the nanojunctions, we can expect to reduce the size by 10 times (assuming the 10 junctions will be equivalent to 1 transistor) [11].

The operational speed of the proposed neuroFPAA and the floating gate transistor-based FPAA can be the same. Since the proposed structure reduces the area of the crossbar by 10X and the crossbar will be half of the complete FPAA (the CAB area is not reduced), the power consumption of the whole system is expected to be reduced by 5X.

4 Conclusion

The significance of this work is that this study introduces an efficient reconfigurable platform for large scale neuromorphic system designs by utilizing nanodevices with CMOS devices. By utilizing nanojunction devices, the proposed nueroFPAA can achieve high-density and low-power operations. The proposed routing and function blocks are specifically designed to suit for both large-scale neuromorphic applications and general analog circuit designs.

A preliminary comparison study is also carried out to compare the existing floating gate-based FPAA and the proposed neuroFPAA. The results demonstrate that by replacing floating gate devices with nanojunction, the density and power can be significantly reduced. However the reproducibility and stability of the nanojunction devices are worse than those of the floating gate transistors. Significant progress is expected to advance the fabrication of nanojunction to enable the proposed neuroFPAA into reality.

References

1. Hall, T.S., Twigg, C.M., Hasler, P.: Large-scale field- programmable analog arrays for analog signal processing. IEEE Trans. Circuits and Systems-I 52(11), 2298–2307 (2005)
2. Gulak, P.G.: Field-programmable analog arrays: past, present and future perspectives. In: IEEE Int'l Conf. on Microelectronics and VLSI, pp. 123–126 (November 1995)
3. Gray, J.D., Twigg, C.M., Hasler, P.: Characteristics and programming of floating-gate pFET switches in an FPAA crossbar network. In: ISCAS 2005, vol. 1, pp. 468–471 (May 2005)
4. Hasler, P.E., Twigg, C.M.: An OTA-based Large-Scale Field Programmable Analog Array (FPAA) for faster On-Chip Communication and Computation. In: ISCAS 2007, pp. 177–180 (May 2007)
5. Tu, D., Liu, M., Wang, W., Haruehanroengra, S.: 3D CMOL: A 3D FPGA using CMOS/nanomaterial hybrid digital circuits. IEE Micro and Nano Letters 2(2), 40–45 (2007)
6. Türel, Ö., Lee, J.H., Ma, X., Likharev, K.K.: Nanoelectronic neuromorphic networks (crossnets): new results. In: Proc. IJCNN 2004, pp. 389–394 (2004)
7. Türel, Ö., Lee, J.H., Ma, X., Likharev, K.K.: Neuromorphic Architectures for Nanoelectronic Circuits. Int. J. of Circuit Theory and Applications 32, 277–302 (2004)
8. Lee, J.H., Likharev, K.K.: Defect-Tolerant nanoelectronic pattern classifiers. Int. J. of Circuit Theory and Applications 35, 239–264 (2007)
9. Gao, C., Hammerstrom, D.: Cortical Models Onto CMOL and CMOL-Architectures and Performance/Price. IEEE Trans. Circuit and System I 54(11) (November 2007)
10. Chen, D., Wang, W., Haruehanroengra, S.: Efficient structures for CMOL circuits. IEE Micro and Nano Letters 1(2), 74–78 (2006)
11. Dong, C., Liu, D., Haruehanroengra, S., Wang, W.: 3D nFPGA: A reconfigurable architecture for 3D CMOS/nanomaterial hybrid digital circuits. IEEE Trans. Circuits and Systems I 54(1), 2489–2501 (2007)
12. Guan, W., Liu, M., Wang, W.: Nonpolar nonvolatile resistive switching in Cu doped ZrO2. IEEE Electronic Device Letters 29(5), 434–438 (2008)

Exploring Multi-layer Graphene Nanoribbon Interconnects

Sansiri Tanachutiwat and Wei Wang

College of Nanoscale Science and Engineering,
State University of New York at Albany, 255 Fuller Rd., Albany, NY, USA
{stanachutiwat,wwang}@uamail.albany.edu

Abstract. In this paper, an improvement of the existing conductance model for the single-layer GNR and a novel conductance model for multi-layer GNR are introduced. The models leverage the recent theoretical and theoretical results, providing consistent conductance/resistance estimations with the experimental results. Using these models, comparison of the resistance of multi-layer GNR with Cu and CNT bundle for the same aspect ratio is carried out. The results demonstrate that multi-layer GNR will be a superior interconnect solution over Cu for 45nm or less technology nodes. This work introduced a promising graphene interconnect by utilizing multiple layers. This might lead to future breakthrough of the new emerging interconnect solution.

Keywords: Graphene, Interconnect, Conductance, Modeling.

1 Introduction

Graphene nanoribbon (GNR) is essentially a monolayer of graphite patterned into a narrow strip, which expected to share many fascinating electrical, mechanical and thermal properties with graphene and carbon nanotube (CNT). Similar to CNT, GNR has a ballistic transport for a very long MFP (in micrometer range) and conduct a large current density in the same order of magnitude reported by CNT [1-5]. The main advantage of GNR over CNT is the lithographic pattern-ability. The chirality of CNT is random during fabrication and the conventional planar process can not be used. But GNR can be fabricated using lithographic pattern technology and its chirality can be tuned by the orientation of the edge termination. Thus, metallic and semiconducting GNR can be controlled to build interconnects and devices respectively, which offers the promise of a large integration of graphene-based interconnects and devices to build future integrated circuits.

In order to provide an efficient on-chip interconnect solution, the multi-layer GNR needs to be considered, which might potentially lower the overall resistance. The stack of GNR layers can be formed during almost all growth processes. Each layer will have around 0.34nm distance due to van de Waal effect.

In this paper, we analyzed the performance of such multi-layer GNRs considering chirality, width, Fermi level shift and effect of electron scattering at the edges. Our

M. Cheng (Ed.): NanoNet 2008, LNICST 3, pp. 49–53, 2009.
© ICST Institute for Computer Sciences, Social Informatics and Telecommunications Engineering 2009

results show that multi-layer GNRs can outperform copper and carbon nanotube in terms of resistance, leading to a promising interconnect solution.

2   Conductance Modeling of Multi-layer GNR

2.1   Single-Layer GNR Conductance

The conductance of a GNR layer depends on the number of the quantum channels and the electron mean free path. It can be calculated as [8]:

$$G = \frac{G_Q \lambda}{L} \,, \tag{1}$$

where G_Q, λ and L are the quantum conductance of graphene ($G_Q = 4e^2/h$ [5,8-9]), the electron mean free path (MFP) considering N_{ch} quantum channels and the length of the GNR, respectively. Since G_Q is a constant and L is a predefined value, we can calculate the conductance G by determining the values of N_{ch} and λ. The calculation of these two values can be found in [6], which are based on: 1) effect of electron scattering, 2) width confinement, 3) GNR chirality, and 4) Fermi level shift. However, the intrinsic and extrinsic phonon scattering effects are not included. Recent results demonstrated that these two phonon effects have a significant impact on λ [10-12].

The results in [5,11] demonstrated that the scattering of intrinsic acoustic phonons (AP) and extrinsic remote interfacial phonons (RIP) due to SiC or SiO$_2$ substrates will reduce λ. Based on these results, we have:

$$\lambda^{-1} = \lambda_e^{-1} + \lambda_{AP}^{-1} + \lambda_{RIP}^{-1} \,. \tag{2}$$

where λ_e, λ_{AP} and λ_{RIP} denote MFP due to electron, intrinsic AP, and extrinsic RIP scattering, respectively. The calculation of λ_e is based on the method in [6] with the modification of N_{ch} to include the subbands above Fermi level within a few kT [8].

Electron scattering consists of scattering in the longitudinal direction due to defects λ_{defect} (the first subbands) as well as scattering in the transverse direction due to edge effect (higher subbands). The effect of width confinement is considered in the estimation of $\Delta E = h v_F/2W$ [6]. The effect of GNR chirality will decide the GNR to be metallic or semiconducting. The metallic GNR has the conduction of the first subband, while the semiconducting GNR does not have the subband, leading to a smaller conductance value [6]. Therefore, the electron scattering MFP is:

$$\lambda_e = \lambda_{\text{defect}} + W \sum_{n=1}^{N_{ch}} \sqrt{\left(\frac{N_{ch}}{n}\right)^2 - 1} \,. \tag{3}$$

The effect of Fermi level shift is due to the charge accumulated at the surface between GNR and substrate/dielectric. This E_F shift from the Dirac point will lead to more conduction channels N_{ch}. The MFP values due to the two phone scattering effects, AP scattering λ_{AP} and RIP scattering λ_{RIP}, can be calculated based on the results in [11-12] as:

$$\lambda_{AP} = \frac{h^2 \rho_s v_s^2 v_F^2 l}{\pi^2 D_A^2 k_B T} \text{ and } \lambda_{RIP} = \alpha l V_g^{1.02} \left(e^{E_0/kT} - 1 \right) , \tag{4}$$

where $v_F = 10^6$ m/s is the Fermi velocity, $v_s = 2.1 \times 10^4$ m/s is the sound velocity, $D_A = 17 \pm 1$ eV [11] is the acoustic deformation potential, and $\rho_s = 6.5 \times 10^{-7}$ kg/m^2 11 is the 2D mass density of graphene. For SiO$_2$ substrate, $\alpha = 0.306$ V$^{-1.02}$ and $E_0 = 104$ meV are obtained from experimental results [11-12]. For the SiC substrate, these parameters will be different and can be found in [12]. Note that both λ_{AP} and λ_{RIP} are temperature dependant. Also, since $\lambda_{RIP} \ll \lambda_{AP}$, the effect of RIP is more significant than that of AP.

By using (2)-(4), we can estimate the conductance of various single-layer GNRs. The estimation results based on the proposed model are more consistent with the measured results of [1, 4] than the results using [6], demonstrating the efficiency of the proposed method.

2.2 Multi-layer GNR Conductance

Based on the above conductance model of the single-layer GNR, we now derive the conductance model of the multi-layer GNR. The number of GNR layer can be consider as $M = \lfloor AR \cdot W/\delta \rfloor$, where AR is the aspect ratio of the interconnects defined in [13] and $\delta \approx 0.34$nm is the inter-layer distance. We have $i = 1, 2, \ldots, M$ to identify each layer in an M-layer GNR.

The effect of charges accumulated at the substrate surface will cause the Fermi level shift. This effective Fermi level shift is observed to decrease exponentially as the layer is away from the substrate, and becomes negligible after the fifth layer [5]. Therefore, for the ith layer and $i \leq 4$, the Fermi level E_F will decay by $e^{-\delta i/\beta}$, where $\beta = 0.387$nm obtained from the fitting curve in [5]. This decay factor is similar to the screening length effect of metal materials reported in [14]. Besides the Fermi level shift, the RIP scattering will disappear when the layer is away from the substrate/dielectric. This effect has been experimentally observed such that it can be neglected after the second layer [11].

Based on the above analysis, we can get G_i for each layer by using $E_{F,i}$ in (3)-(4). Then, we obtain the total conductance of an M-layer GNR as:

$$G_{GNR} = \sum_{i=1}^{M} G_i , \tag{5}$$

In order to demonstrate the efficiency of the proposed model (5), we estimate the conductance of a 6-layer GNR [5]. The simulation result matches the experiment result.

3 Conductance Comparison

Using the proposed multi-layer GNR conductance models, we estimate the conductance of various GNR interconnect solutions and compare the results with copper [13], monolayer SWCNT and mixed CNT bundle [8] as shown in Fig. 1. It is seen in Fig. 1 that the single-layer GNR has a higher resistance than CNT and Cu wires, which is consistent with the result in [8]. But for the same aspect ratio, the metallic multi-layer GNR has a lower resistance than Cu and the CNT bundle. For the 45nm, 32nm, 22nm and 16nm technology nodes, the improvements of the multi-layer GNR over Cu are 12.8%, 46.1%, 68.8% and 81.4%, respectively. Note that the semiconducting GNR will have a poor conductance performance, which is consistent with the measurement results in [3]. Therefore, in order to utilize the multi-layer GNR, more metallic layers are required to provide an efficient interconnect solution.

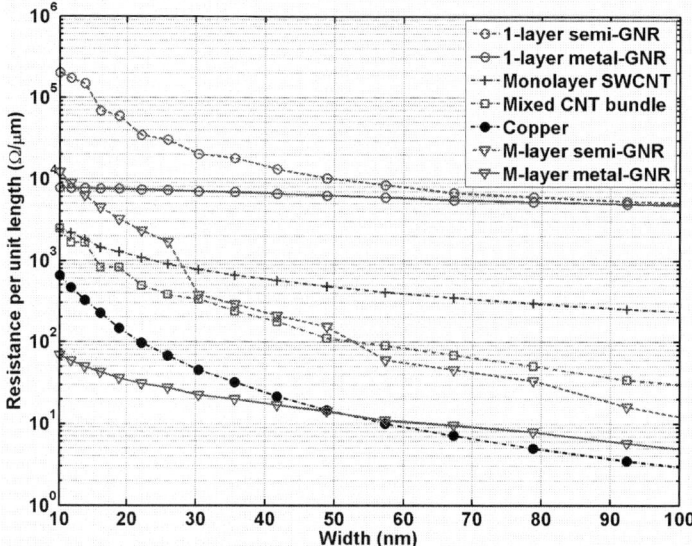

Fig. 1. Comparison of resistance per unit length versus different width of different technology nodes for various interconnect materials. i.e. single-layer GNR obtained from [6], the proposed single-layer GNR, proposed multi-layer GNR (aspect ratios are obtained from [13]), carbon nanotube bundle 8 (aspect ratios are obtained from [13]), single-layer carbon nanotube array [8] and copper interconnect [13].

4 Conclusion

In this paper, we improved the existing conductance models for the single-layer GNR and introduced a novel conductance model for multi-layer GNR. The models leverage the recent theoretical and theoretical results, providing a consistent conductance/resistance estimation with the experimental results. Using these models, we compared the conductance of multi-layer GNR with Cu for the same aspect ratio. The results demonstrate that multi-layer GNR will be a superior interconnect solution over Cu for 45nm or less technology nodes.

The novelty of this work is that the new conductance model of GNR considers the realistic substrate/dielectric and layer effect. The results leading to a promising multi-layer graphene interconnect solution. The significance is that the introduction the multi-layer GNR interconnect concepts and models might lead to future breakthrough of the new emerging interconnect solution.

Acknowledgment

This work has been supported in part by the AFSTTR and MARCO (via IFC Center). Useful discussions with Ji Ung Lee and Robert Geer are gratefully acknowledged.

References

1. Berger, C., et al.: Electronic Confinement and Coherence in Patterned Epitaxial Graphene. Science 312(5777), 1191–1196 (2006)
2. Hass, J., et al.: Why Multilayer Graphene on 4H-SiC(0001̄) Behaves Like a Single Sheet of Graphene. Phys. Rev. Lett. 100, 125504 (2008)
3. Han, M.Y., Zyilmaz, B.O., Zhang, Y., Kim, P.: Energy Band-Gap Engineering of Graphene Nanoribbons. Phys. Rev. Lett. 98, 206805 (2007)
4. Chen, Z., Lin, Y., Rooks, M.J., Avouris, P.: Graphene nano-ribbon electronics. Physica E 40(2), 228–232 (2007)
5. Wang, H.M., Wu, Y.H., Ni, Z.H., Shen, Z.X.: Electronic transport and layer engineering in multilayer graphene structures. Applied Physics Letters 92, 03504 (2008)
6. Naeemi, Meindl, J.D.: Conductance Modeling for Graphene Nanoribbon (GNR) Interconnects. IEEE Electron Device Letters 28(5), 428–431 (2007)
7. Cserti, J., Csordás, A., Dávid, G.: Role of the Trigonal Warping on the Minimal Conductivity of Bilayer Graphene. Phys. Rev. Lett. 99, 066802 (2007)
8. Haruehanroengra, S., Wang, W.: Analyzing Conductance of Mixed Carbon-Nanotube Bundles for Interconnect Applications. IEEE EDL 28(8), 756–759 (2007)
9. Latil, S., Henrard, L.: Charge Carriers in Few-Layer Graphene Films. Phys. Rev. Lett. 97(3), 036803 (2006)
10. Jang, S.A., Chen, J.H., Williams, E.D., Sarma, S.D., Fuhrer, M.S.: Tuning the Effective Fine Structure Constant in Graphene: Opposing Effects of Dielectric Screening on Short- and Long-Range Potential Scattering, http://arxiv.org/abs/0805.3780v1
11. Chen, J.H., Jang, C., Xiao, S., Ishigami, M., Fuhrer, M.S.: Intrinsic and Extrinsic Performance Limits of Graphene Devices on SiO2. Nature Nanotech, 206–209 (2008)
12. Fratini, S., Guinea, F.: Substrate-Limited Electron Dynamics in Graphene. Phys. Rev. B 77, 195415 (2008)
13. International Technology Roadmap for Semiconductor (ITRS 2007) (2007), http://public.itrs.net
14. Ohta, T., et al.: Interlayer Interaction and Electronic Screening in Multilayer Graphene Investigated with Angle-Resolved Photoemission Spectroscopy. Phys. Rev. Lett. 98, 206802 (2007)

Digital Microfluidic Logic Gates

Yang Zhao, Tao Xu, and Krishnendu Chakrabarty

Department of Electrical and Computer Engineering,
Duke University, Durham, NC 27708, USA
{yz61,tx,krish}@ee.duke.edu

Abstract. Microfluidic computing is an emerging application for microfluidics technology. We propose microfluidic logic gates based on digital microfluidics. Using the principle of electrowetting-on-dielectric, AND, OR, NOT and XOR gates are implemented through basic droplet-handling operations such as transporting, merging and splitting. The same input-output interpretation enables the cascading of gates to create nontrivial computing systems. We present a potential application for microfluidic logic gates by implementing microfluidic logic operations for on-chip HIV test.

Keywords: microfluidic computing, digital microfluidics, logic gate.

1 Introduction

Microfluidics technology has made great strides in recent years [1]. The applications of this emerging technology include immunoassays, environmental toxicity monitoring and high through-put DNA sequencing. An especially promising technology platform is based on the principle of electrowetting-on-dielectric. Discrete droplets of nanoliter volumes can be manipulated in a "digital" manner on a two-dimensional electrode array. Hence this technology is referred to as "digital microfluidics" [2].

An especially promising application is the use of droplets for microfluidic computing. Microfluidic computing inherits the advantages of both microfluidics for sensing and computing for information processing [3]. The speed of microfluidic computing is much less than silicon-based computing devices. Hence microfluidic computing will not replace conventional computing devices; nevertheless, it will enhance microfluidic technology through direct incorporation of computing functions on-chip with other primary sensing functions. Microfluidic computing devices can be implemented in various ways, such as electrochemical reactions [4], relative resistance [5], bubbles in electronic channels [6]. However, a drawback of these methods is that they assign different interpretations to inputs and outputs, which makes cascading of gates difficult.

We propose logic gates based on digital microfluidics. We implement AND, OR, NOT and XOR gates through basic droplet-handling operations such as transportation, merging, and splitting by using the principle of electrowetting-on-dielectric. Nontrivial computing systems can be created by cascading the microfluidic logic gates that have the same input-output interpretation. A potential

M. Cheng (Ed.): NanoNet 2008, LNICST 3, pp. 54–60, 2009.

application for microfluidic logic gates is presented by implementing microfluidic logic operations for on-chip HIV test.

2   Digital Microfluidic Platform

In digital microfluidics, droplets of nanoliter volumes are manipulated on a two-dimensional electrode array [1]. A unit cell in the array includes a pair of electrodes that acts as two parallel plates. The bottom plate contains a patterned array of individually controlled electrodes, and the top plate is coated with a continuous ground electrode. All electrodes are formed by optically transparent indium tin oxide (ITO). A dielectric insulator, i.e., parylene C, coated with a hydrophobic film of Teflon AF, is added to the top and bottom plates to decrease the wettability of the surface and to add capacitance between the droplet and the control electrode. A droplet rests on a hydrophobic surface over an electrode.

Droplets are moved by applying a control voltage to a unit cell adjacent to the droplet and, at the same time, deactivating the cell just under the droplet. This electronic method of wettability control creates interfacial tension gradients that move the droplets to the charged electrode. Fluid-handling operations such as droplet merging, splitting, mixing, and dispensing can be executed in a similar manner. Droplet routes and operation schedules are programmed into a microcontroller that drives the electrodes. Design automation techniques for digital microfluidics are now being developed [10].

3   Digital Microfluidic Logic Gates

In the digital microfluidic platform, droplets of unit volume (1x) or larger can be easily moved [7]. A droplet of 0.5x volume is not large enough to have sufficient overlap with an adjacent electrode; hence it cannot be moved [7]. It has been verified experimentally that the times required for dispensing one droplet, splitting a droplet into two, merging two droplets into one, and transporting a droplet to an adjacent electrode are nearly identical. This duration is defined as one clock cycle.

The definitions for logic values '0' or '1' are as follows: the presence of a droplet of 1x volume at an input or output port indicates a logic value of '1'. The absence of a droplet at an input or output port indicates the logic value '0'. The same interpretations at inputs and outputs enable the output of one gate to be fed as an input to another gate, thus logic gates can be easily cascaded.

Fig. 1 shows the schematics of the 2-input OR, 2-input AND, NOT, and XOR gates. The OR gate in Fig. 1(a) incorporates a waste reservoir (WR) and twelve indexed electrodes. Electrode 1 and Electrode 2 are the two input ports X_1 and X_2; Electrode 3 is the reference port (R), from which one reference droplet is injected into the OR gate. Electrode 9 is the output port (Z) where a detector can be placed to determine the output logic value of the OR gate. Such detections to indicate the presence or absence of a droplet can be easily implemented using capacitive measurements [11]. Electrode 12 is the washing

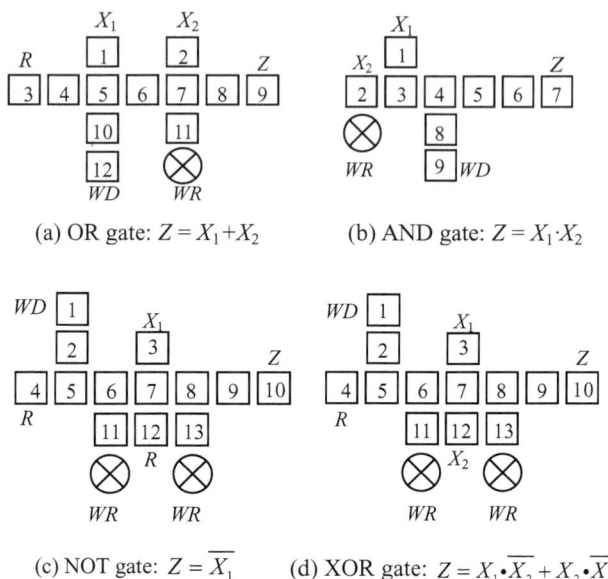

(a) OR gate: $Z = X_1 + X_2$ (b) AND gate: $Z = X_1 \cdot X_2$

(c) NOT gate: $Z = \overline{X_1}$ (d) XOR gate: $Z = X_1 \cdot \overline{X_2} + X_2 \cdot \overline{X_1}$

Fig. 1. Schematics of microfluidic logic gates

Table 1. Actuation-voltage sequence for the OR gate

Clock	Electrode No.											
cycle	1	2	3	4	5	6	7	8	9	10	11	12
0	1	1	1	0	0	F	0	F	F	0	F	1
1	0	0	1	0	1	0	1	0	F	0	0	1
2	0	0	1	0	0	1	0	0	F	0	0	1
3	0	0	1	0	1	0	1	0	F	0	0	1
4	0	0	0	1	0	0	0	0	F	0	1	1
5	0	F	0	0	1	0	0	F	F	0	0	1
6	0	F	F	0	0	1	0	F	F	0	F	1
7	F	0	F	F	0	0	1	0	F	0	0	1
8	F	0	F	F	0	1	0	1	0	0	0	1
9	F	F	F	F	0	1	0	0	1	0	F	1

port (WD), from which a washing droplet is injected after the logic operation to collect the residual droplets and move them to the waste reservoir. The sequence of control voltage applied to each electrode is shown in Table 1. A '1' ('0') entry in the table indicates a high (low) voltage to the corresponding electrode in that clock cycle. An 'F' entry indicates a floating signal, i.e., it is not required to be either active or inactive. The sequence of control voltages is independent of the input logic values. Fig. 2 describes the cycle-by-cycle operation of the OR gate for $X_1 X_2 = 11$.

The delay of the OR gate is 9 clock cycles, independent of the inputs. At the beginning of clock cycle 10, the droplet on the washing port (Electrode 12) is

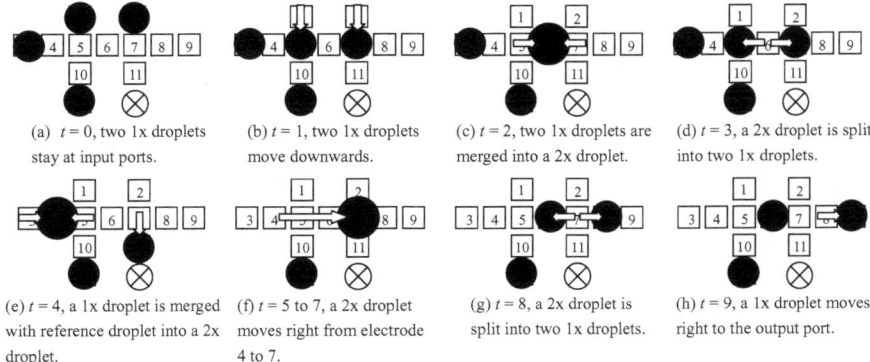

(a) $t = 0$, two 1x droplets stay at input ports.

(b) $t = 1$, two 1x droplets move downwards.

(c) $t = 2$, two 1x droplets are merged into a 2x droplet.

(d) $t = 3$, a 2x droplet is split into two 1x droplets.

(e) $t = 4$, a 1x droplet is merged with reference droplet into a 2x droplet.

(f) $t = 5$ to 7, a 2x droplet moves right from electrode 4 to 7.

(g) $t = 8$, a 2x droplet is split into two 1x droplets.

(h) $t = 9$, a 1x droplet moves right to the output port.

Fig. 2. Operation of the OR gate for input 11

routed to merge with the residual droplets, and the result is transported to the waste reservoir. After this washing process, no droplet is left on the electrodes, and this gate is clean for the next operation.

To experimentally verify the OR gate, we configured it on a fabricated lab-on-chip, then activated the corresponding electrodes to perform on-chip cycle-by-cycle operations in the laboratory. In this experiment, we use a lab-on-chip with an electrode pitch of 1.5 mm and a gap spacing of 0.475 mm. The droplets are dispensed from the on-chip reservoirs that are filled by DI water with black dye. The voltage set-up for the splitting process is 250 V (input voltage for PCB). Under this voltage set-up, the droplet with volume equal to or larger than 1x can be split into two droplets with equal volumes. The voltage set-up for transportation is in the range of 80 V to 90 V (input voltage for PCB). Under this voltage set-up, only droplets with volume equal to or larger than 1x can be

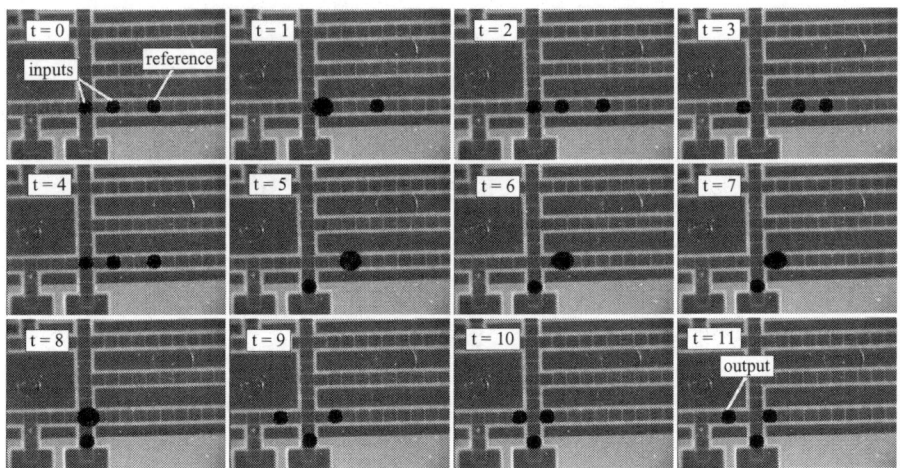

Fig. 3. On-chip cycle-by-cycle operation for the OR gate with input 11

moved to the adjacent activated electrode. Fig. 3 shows the operation of the gate for $X_1 X_2 = 11$. When $t = 0$, two 1x droplets stay on the electrodes representing two inputs, while one 1x droplet stays on the electrode representing the reference port. Operations from $t = 1$ to 10 are the same as that in Fig. 2. Note that the splitting step in the experiment occupies five electrodes and lasts for three clock cycles, as shown from $t = 2$ to 4. This is because we want to ensure even and thorough splitting, to acquire two split droplets with equal volume. At $t = 11$, there is one 1x droplet on the electrode representing the output. Experimental results demonstrate the feasibility of the OR gate for different input values.

Fig. 1(b) illustrates the schematic of a 2-input microfluidic AND gate. The sequence of control voltages applied to each electrode is shown in Table 2. The delay of the AND gate is 5 clock cycles. Fig. 1(c) shows a microfluidic inverter. The sequence of actuation voltages applied on each electrode is shown in Table 2. The delay of the inverter is 9 clock cycles. Fig. 1(d) illustrates the schematic of a microfluidic XOR gate. The logic function of this XOR gate is as follows: $Z = X_1 \cdot \overline{X_2} + X_2 \cdot \overline{X_1}$. The sequence of control voltages applied to each electrode of the XOR gate is the same as that of the inverter. The delay of this XOR gate is also 9 clock cycles.

Table 2. Actuation voltage sequence for the AND gate and the inverter (the values for the inverter are shown in parentheses)

Clock cycle	Electrode No.												
	1	2	3	4	5	6	7	8	9	10	11	12	13
0	1(1)	1(0)	0(1)	F(1)	F(0)	F(F)	F(0)	0(F)	1(F)	(F)	(0)	(1)	(0)
1	0(1)	0(0)	1(0)	0(1)	F(0)	F(0)	F(1)	0(0)	1(F)	(F)	(0)	(0)	(0)
2	0(1)	1(0)	0(0)	1(1)	0(0)	F(1)	F(0)	0(1)	1(0)	(F)	(0)	(0)	(0)
3	F(1)	0(0)	0(F)	0(1)	1(0)	0(0)	F(0)	0(0)	1(0)	(F)	(1)	(0)	(1)
4	F(1)	F(0)	F(F)	0(0)	0(1)	1(0)	0(F)	0(0)	1(F)	(F)	(0)	(0)	(0)
5	F(1)	F(0)	F(F)	F(0)	0(0)	0(1)	1(0)	0(F)	1(F)	(F)	(0)	(F)	(F)
6	(1)	(0)	(0)	(F)	(0)	(0)	(1)	(0)	(F)	(F)	(0)	(0)	(F)
7	(1)	(0)	(0)	(F)	(F)	(0)	(0)	(1)	(0)	(F)	(F)	(0)	(0)
8	(1)	(0)	(0)	(F)	(F)	(0)	(1)	(0)	(1)	(0)	(F)	(0)	(0)
9	(1)	(0)	(0)	(F)	(F)	(0)	(1)	(0)	(0)	(1)	(F)	(0)	(F)

4   Potential Application of Microfluidic Logic

HIV test is integral to HIV prevention, treatment, and care efforts. The knowledge of an individual's HIV status is important for preventing the spread of this disease. Early knowledge of HIV status is also important for offering those who are HIV positive with medical care and services to improve quality of life [9]. Early diagnosis is important for effective treatments and to prevent transmission of HIV infection to other individuals.

A common test for HIV can be implemented on a digital microfluidic platform. Tests used for the diagnosis of HIV infection should achieve a high degree of both sensitivity and specificity. This is achieved by combining the screening assay and the confirmatory assay. In the US, the enzyme-linked immunosorbent assays (ELISA) are commonly used for screening to look for antibodies to HIV, and the Western Blot procedure is commonly used for confirmatory assay. A positive ELISA must be used with a follow-up (confirmatory) test such as the Western Blot to make a positive diagnosis [8].

The ELISA test, on blood drawn from a vein, is the most common screening test used to look for antibodies to HIV. The results of the ELISA test, which are in the form of chromogenic, fluorogenic, or electrochemical signals, can be viewed by optical or electrochemical devices. The results of the ELISA test are qualitative, either positive or negative. If the result of the ELISA test is negative, there are no HIV antibodies in the blood sample, confirming that the individual is not infected with HIV. If the result of ELISA is positive, a follow-up (confirmatory) test such as the Western Blot must be used to make a positive diagnosis.

The results of Western Blot are analyzed in the following way: If no viral bands are detected, the result is negative. If at least one viral band is present for each gene-product group, the result is positive. Tests in which less than the required number of viral bands is detected are defined as indeterminate. The person who has the indeterminate result should be retested, as later tests may be more conclusive.

The microfluidic logic gates can be used to implement logic operations for the on-chip HIV test. Although the speed of microfluidic computing using microfluidic logic gates is much less than silicon-based computing devices, the microfluidic logic operations will enhance the on-chip HIV test through direct incorporation of computing functions on-chip with primary HIV test operations without the conversion between fluidic signals and electrical signals.

One logic operation is outlined here to evaluate the results of Western Blot. Assuming there are n gene-product groups (typically $n = 3$), we define $X_j = 1$ if there is at least one viral band for gene-product group j. We use two logic functions as follows.

$$Y = OR(X_1, X_2, \ldots, X_j, \ldots, X_n),$$
$$Z = AND(X_1, X_2, \ldots, X_j, \ldots, X_n).$$

If no viral bands are detected, then $Y = 0$, else $Y = 1$. If at least one viral band for each of the n gene-product groups is present, then $Z = 1$, else $Z = 0$. The results of Western Blot can be evaluated by the combination of Y and Z, as shown in Table 3.

Table 3. Result-evaluation table of Western Blot

Signal Y	Signal Z	Western Blot Result
0	0	negative
0	1	forbidden state
1	0	indeterminate
1	1	positive

5 Conclusions

Microfluidic logic gates have been proposed based on the digital microfluidic platform. Microfluidic AND, OR, NOT and XOR gates can be implemented through
basic droplet-handling operations such as transporting, merging and splitting by
using the principle of electrowetting-on-dielectric. Experimental results demonstrate the feasibility of the microfluidic logic gates for different input values.
A potential application for microfluidic logic gates is to implement microfluidic
logic operations for on-chip HIV test.

References

1. Fair, R.B., et al.: Chemical and Biological Applications of Digital-Microfluidic Devices. IEEE Design & Test of Computers 24, 10–24 (2007)
2. Chakrabarty, K., Su, F.: Digital Microfluidic Biochips: Synthesis, Testing, and
   Reconfiguration Techniques. CRC Press, Boca Raton (2006)
3. Marr, D.W.M., Munakata, T.: Micro/Nanofluidic computing. Comm. ACM 50,
   64–68 (2007)
4. Zhan, W., Crooks, R.M.: Microelectrochemical logic circuits. J. Am. Chem.
   Soc. 125(33), 9934–9935 (2003)
5. Vestad, T., et al.: Flow resistance for microfluidic logic operations. Applied Physics
   Letters 84(25), 5074–5075 (2004)
6. Prakash, M., Gershenfeld, N.: Microfluidic bubble logic. Science 315(5813), 832–835
   (2007)
7. Pollack, M.G., et al.: Electrowetting-based actuation of droplets for integrated
   microfluidics. Lab on a Chip 2(1), 96–101 (2002)
8. National HIV and STD Testing Resources, http://www.hivtest.org
9. Janssen, R.S., et al.: Advancing HIV prevention: New strategies for a changing
   epidemic. Morbidity and Mortality Weekly Report 52(15), 329–332 (2003)
10. Xu, T., Chakrabarty, K.: Integrated droplet routing in the synthesis of microfluidic
    biochips. In: Proc. DAC, pp. 948–953 (2007)
11. Xu, T., Chakrabarty, K.: Parallel scan-like test and multiple-defect diagnosis for
    digital microfluidic biochips. IEEE Transactions on Biomedical Circuits and Sys-
    tems 1, 148–158 (2007)

Application of Molecular Electronics Devices in Digital Circuit Design

Ci Lei[1], Dinesh Pamunuwa[1], Steven Bailey[2], and Colin Lambert[2]

[1] Engineering Department, Lancaster University, Lancaster, LA1 4YW, UK
`c.lei@lancaster.ac.uk`
[2] Physics Department, Lancaster University, Lancaster, LA1 4YW, UK

Abstract. The Breit-Wigner resonance formula is used to model a class of molecular electronic devices, in order to establish an abstract model for exploration of their applicability in future nanoelectronic systems. The model is used to characterize molecular device I-V curves in terms of the coupling between the molecule and the leads, and demonstrate digital circuit functionality. Circuit metrics such as noise margin, speed and power are investigated.

Keywords: molecular electronics, circuit simulation, nanotechnology.

1 Introduction

Due to recent success in measuring the $I - V$ characteristics of individual or small groups of molecules, developing and controlling electronic molecules that can serve as the active elements in future nano-electronic circuits has become a current objective in this field [1,2,3,4]. Hence developing a hierarchy of device and interconnect models and an efficient simulation methodology is a key step in exploring the system capabilities of molecular device based circuits, as part of the search for miniaturization of devices beyond CMOS, as predicted by the ITRS[5]. In this paper, we propose a refinement of a compact model introduced by Purdue University [6,7,8] that describes the static or steady-state behaviour of a device. The extension allows the dynamic behavior of the device to be modeled and hence investigation of the potential performance impact that general molecular electronics device technology could have on Very Large Scale Integration (VLSI) circuit applications.

2 Device Model

Static Behavior. The transmitted current I through the metal-molecule-metal junction is proportional to the transmission probability $T(E)$ describing the ease with which electrons can scatter through the molecule from the source lead into the drain lead for a range of energy levels around the Fermi energy of the leads. From [9], the current is computed using

M. Cheng (Ed.): NanoNet 2008, LNICST 3, pp. 61–65, 2009.

$$I(V) = \frac{2e}{h} \int_{-\infty}^{\infty} T(E)\left(\frac{1}{\exp[E-\mu_s]/kT+1} - \frac{1}{\exp[E-\mu_d]/kT+1}\right)dE \quad (1)$$

where e is the electron charge and h is Planck's constant. For symmetric molecules, the two electro-chemical potentials μ_s and μ_d (refering to the source and drain respectively) are defined to be $\mu_s = E_f + eV/2$ and $\mu_d = E_f - eV/2$, where V is the bias voltage applied between the source and drain, and E_f is the Fermi energy for the contacts. In the case of weak interaction between the leads and the molecule, it is well known [10] that the transmission probability can be approximated by the Breit-Wigner formula $T(E) = \frac{4\Gamma_1\Gamma_2}{(E-\epsilon_0)^2+(\Gamma_1+\Gamma_2)^2}$ where the variables Γ_1 and Γ_2 represent the broadenings of molecular levels by hybridization with the contacts and the molecular level ϵ_0 is related to the intrinsic chemistry of the molecule. A plot of this probability and corresponding I-V characteristics calculated using Eq.(1) is given in Fig.(1).

Dynamic Behavior. For the purpose of circuit performance investigation, knowledge of an equivalent time constant related to the physical description of the device is sufficient for its dynamic characterization. This time constant can be estimated by calculating the difference between the time spent by the electron in the region of scattering interaction and the time spent in the same region in the absence of scattering interaction.

The event of scattering is described using the scattering matrix $S(E)$ [11]. $S(E)$ is unitary and can be written as $S(E) = \exp(2i\delta)$, where $\delta(E) = -i \log S(E)$ is called the scattering phase shift. For narrow isolated resonance with a single open channel, $\delta(E)$ may be parameterized in terms of the resonance position ϵ_0 and width $\Gamma_1 + \Gamma_2$ by the Breit-Wigner one-level formula $\tan[\delta(E)] = -\frac{(\Gamma_1+\Gamma_2)/2}{E-\epsilon_0}$ [10]. The definition of the phase shift $\delta(E)$ has a very intuitive physical meaning, as it is related to the Wigner delay time τ which is the additional time spent in the scattering process compared to free motion by $\tau \times |\frac{(\Gamma_1+\Gamma_2)}{E-\epsilon_0+i(\Gamma_1+\Gamma_2)/2}|^2 = 2\hbar(d\delta/dE)$. Hence under the weak coupling condition, the time constant of the

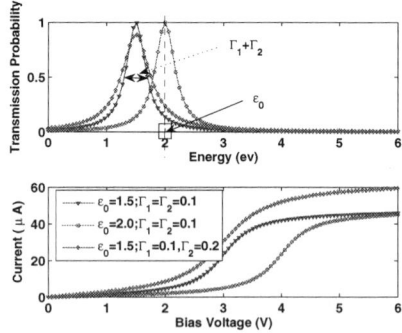

Fig. 1. The transmission probability as a function of molecular energy and corresponding $I - V$ characteristics for different coupling parameters

molecular device is expressed as $\tau = \frac{\hbar}{\Gamma_1 + \Gamma_2}$ at $E = \epsilon_0$. We propose to use this time constant to describe the dynamic behavior of the device as an exponential decay; i.e. τ is the time taken to reach 63.2% of the final value of the step response of the device. The current through the device as a function of time can then be described as $I(t) = I_{sat}(1 - \exp(-t/\tau))$, where I_{sat} is the pulse amplitude calculated using the static model for a given voltage.

3   Implementation and Results

The model proposed in Sec.2 can be used to describe two terminal devices using resonance tunneling as a conduction mechanism. Such a device can be used as a two-terminal switch with the "on" or "off" states being a function of the terminal voltage. The device models have been implemented using the analog hardware description language VerilogA, allowing arbitrary circuit configurations to be simulated for static and dynamic behavior within the analog and mixed-signal simulation environment from Cadence that includes the Spectre circuit simulator.

Fig.(2a) shows the generic circuit arrangement for an individual device, the simplest arrangement that suffices for illustration of circuit operation. A 3-D plot that shows the voltage across the device and the current through the device for different values of R is shown in Fig.(2b). Operating points for different values of R are shown in black on the plot, where $\Gamma_1 = \Gamma_2 = 0.1$ and $\epsilon_0 = 1.5$. The value of the pull-up (and pull-down) resistors in a more complex circuit will thus determine the DC bias point of the molecular device, and establish the regime of operation in which the device operates. Shown in Fig.(2c) is a three-input AND-OR gate implemented in a cross-bar architecture [12] and the DC simulation of V_{out} for various input combinations. The connectivity between nodes is established by molecular devices, while the horizontal and vertical lines represent nanowires.

The inputs are labeled alphabetically. For this logic gate, a low input is held at ground, and a high input is held at $+1V$. The schematic shows pull-up and

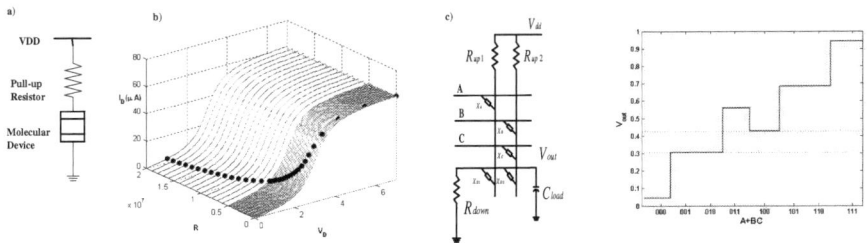

Fig. 2. *a*) A generic cross-bar circuit that utilizes a molecular device; *b*) black dots show how the DC operating point changes with the value of the biasing resistor; *c*) Circuit layout for the AND-OR gate; Derived truth table

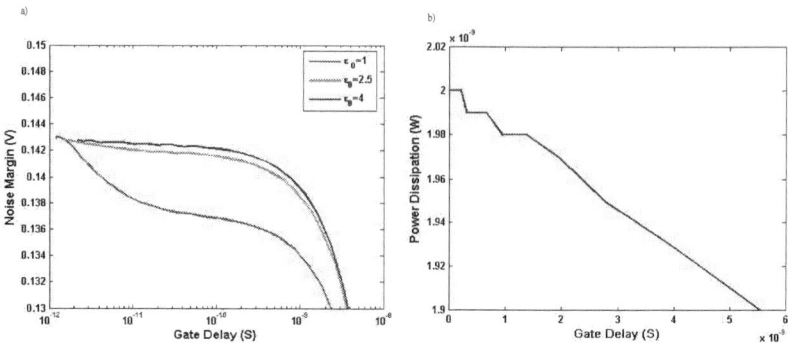

Fig. 3. *a*) The relationship between the speed and the noise margin of the gate for different values of ϵ; *b*) Trade-off between the power dissipation and speed of the gate

pull-down resistors labeled R_{up1}, R_{up2} and R_{down}. The gate uses five molecular devices X_A, X_B, X_C, X_{D1}, and X_{D2}. In the first instance we set the coupling values for all devices as 0.1 while $R_{up1} = R_{up2} = R_{down} = 1G$ and $C_{load} = 100aF$. The plot in Fig.(2c) is a truth table where the output is given as a voltage level. The output voltages of $V_{out} = 0.42V$ and $V_{out} = 0.56V$ are interpreted as binary "1" while voltages less than $0.3V$ are interpreted as logic "0". It can be seen that the circuit acts as a three-input AND-OR gate.

One of the most important metrics of interest in characterizing static behavior is the DC noise margin, which is a function of the separation between the high and low voltage levels, and is defined as $(V_{high} - V_{low})$ here for convenience. As mentioned, the delay time of the functional gate is a function of the coupling parameters. Hence a relationship can be established between the noise margin and the rise time of the step response. This relationship is shown in Fig.(3a).

Power dissipation results in heating that could have a particularly negative cumulative impact in an ultra-dense molecular electronic system with many closely-spaced wires, switches, gates, and functions. It is also crucial to estimate the power for a system that has a complexity similar to that of of a representative digital system such as a processor. The static and dynamic models proposed in this paper and our simulation methodology allows the calculation of such quantities. The gates constructed in this PLA type architecture have static power dissipation, on the order of tens to hundreds of nanowatts per complex gate. the I-V curve will influence the power dissipation, and this can be captured by the plot of Fig.(3b), which shows the variation of static power dissipation with gate delay. The static power dissipation will depend on the state the gate is in, and a representative average for this 3-input AND-OR gate is around 2nW. This translates to a total static power consumption of around 0.002W for a system of 1 million gates, which can be up to several orders of magnitude improvement on a state of the art CMOS implementation in a 50nm technology.

4   Conclusion

The Breit-Wigner formula for resonant conductance is used to extract current-voltage curves for a generic molecular electronic device, based on fundamental physical constants related to the electronic structure of the molecule. The Wigner delay time derived from the scattering matrix is used as the time constant to characterize the dynamic behavior of the device. Using the composite model, digital logic behaviour was demonstrated for a molecular-device-based circuit, and a comprehensive analysis of performance metrics of interest including noise margin, speed and power dissipation was carried out. The study reveals a separation between distinct logic levels that is several orders of magnitude greater than the thermal noise induced voltage at room temperature, and an equivalent output impedance of a few M ohms, resulting in a relatively slow (dis)charging time constant for sub fF loads that dominates the internal charging time of the device itself. Due to the logic architecture, the power consumption has an input-pattern-dependent static component(a few nW for the 3-input AND-OR gate considered in the study) in addition to the familiar CMOS type dynamic component, that depends on the load and switching frequency. The general viability of molecular-based electronics for ultra dense low-power applications based purely on their functional capability and performance metrics appears to be promising.

References

1. Reed, M.A.: Molecular-scale electronics. Proc. IEEE 87, 652–658 (1999)
2. Chen, J., Reed, M., Rawlett, A., Tour, J.: Large On-Off Ratios and Negative Differential Resistance in a Molecular Electronic Device. Science 286, 1550–1552 (1999)
3. Bourgoin, J.P.: Molecular electronics: a review of metal-molecule-metal junction. Lect. Notes Phys. 579, 105–124 (2001)
4. Kergueris, C., Bourgoin, J.P., Palacin, S.: Electron transport through a metal-molecule-metal junction. Phys. Rev. B 59, 505–513 (1999)
5. The International Technology Roadmap for Semiconductors (2007), http://www.itrs.net
6. Datta, S.: Electrical resistance:an atomistic view. Nanotechnology 15, s433–s451 (2004)
7. Datta, S.: Quantum Transport: Atom to Transistor. Cambridge University Press, Cambridge (2005)
8. Lei, C., Pamunuwa, D., Bailey, S., Lambert, C.: Molecular Electronics Device Modeling for System Design. Presented at the IEEE Conf. Nanotechnology (IEEE-NANO), HongKong (2007)
9. Landauer, R.: Can a length of perfect conductor have a resistance. Phys. Lett. 85A, 91–93 (1981)
10. Breit, G., Wigner, E.: Capture of Slow Neutrons. Phys. Rev. 49, 519–531 (1936)
11. Bauer, M., Mello, P.A., McVoy, K.W.: Time delay in nuclear reactions. Zeit. Physik A 293, 151–163 (1979)
12. Collier, C.P., Wong, E.W., Belohradsky, M., Raymo, F.M., Stoddart, J.F., Kuekes, P.J., Williams, R.S., Heath, J.R.: Electronically configurable molecular-based logic gates. Science 285, 391–394 (1999)

A Voltage Controlled Nano Addressing Circuit

Bao Liu

University of Texas, San Antonio TX 78249, USA

Abstract. A voltage controlled nano addressing circuit is proposed, which (1) improves yield and enables aggressive scaling with no requirement of precise layout design, (2) achieves precision of addressing by transistor current-to-voltage sensitivity in the circuit and applied external address voltages, and (3) is adaptive to and more robust in the presence of process variations which are expected to be prevalent in nanoelectronic designs.

Keywords: Nanoelectronics, Nano Architecture, Nano Addressing.

An outstanding challenge for realizing nanoelectronic systems is how to precisely address a nanoscale wire in an array for configuration or data IO. The existing nano addressing mechanisms are based on binary decoders to select one of 2^n nanoscale data lines based on n microscale address lines. They are either (1) randomized contact decoders [9], (2) addressing undifferentiated nanoscale wires by (lithography defined) differential microscale wires [4], or (3) addressing differentiated nanoscale wires by undifferentiated microscale wires [1,5]. All these existing nano addressing mechanisms require precise layout design, which is unlikely in nanotechnology, wherein regular structures are expected to grow in bottom-up self-assembly processes [6].

I propose a voltage controlled nano addressing circuit (Fig. 1 (left)), which includes two rows of field effect transistors, of which the source/drain regions are connected to the data lines (nanoscale wires, e.g., carbon nanotubes), while the gates are connected to the address lines (which can be microscale wires or even nanoscale wires). Continuously tunable external voltages (V_{dda1}, V_{ssa1}, V_{dda2}, and V_{ssa2}) are applied to the address lines and the transistor gates. All components in this structure are designed as uniform, e.g., the transistors are identical, and the address lines have uniform serial resistance. The external voltages are applied such that a decreasing and an increasing array of gate voltages are applied to the transistors in the first and the second row, respectively. As a result, the data lines have different conductivity depending on their locations. By applying different addressing voltages (V_{dda1}, V_{ssa1}, V_{dda2}, and V_{ssa2}), this circuit is able to selectively address one of the data lines in the array.

For example, to address a carbon nanotube (CNT) in an array, each nanotube is gated by two N-type CNT field effect transistors (CNFETs) [8] of $6.4nm$ gate width and $32nm$ channel length, as are given by the Stanford CNFET

M. Cheng (Ed.): NanoNet 2008, LNICST 3, pp. 66–68, 2009.

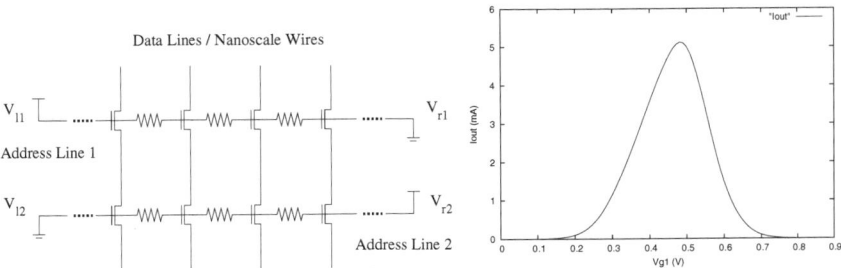

Fig. 1. A voltage controlled nano addressing circuit (left), and nanotube current I_{out} (mA) vs. CNFET gate voltage $V_{g1}(V)$ in the first address line (right)

compact model [2]. The two CNFETs in each nanotube are given a voltage drop of $V_{dd} = 1V$. The external address voltages are $V_{dda1} = V_{dda2} = 1V$, $V_{ssa1} = V_{ssa2} = 0$. SPICE simulation gives the current for each nanotube in the array (data lines) with different gate voltage in the first address line (Fig. 1 (right)). We have the following observations.

1. The nanotubes carry a significant current only with specific gate voltages, allowing addressing of a specific carbon nanotube by applying different addressing voltages.
2. To precisely address a single nanotube, two adjacent nanotubes must carry significant different currents. This can be achieved by (1) increasing the current-to-voltage sensitivity of the transistors [8], or (2) increasing the gate voltage difference between two adjacent nanotubes.
3. Process variations (e.g., of data line shifting and contact resistance) can be cancelled by tuning the addressing voltages, or (e.g., of address line geometry or resistance) have little effect.
4. The proposed circuit requires only uniform components in a regular structure, avoiding precise layout design, which significnatly improves yield and enables aggressive scaling of nanoelectronic systems.

References

1. DeHon, A., Lincoln, P., Savage, J.E.: Stochastic Assembly of Sublithographic Nanoscale Interface. IEEE Trans. Nanotechnology 2(3), 165–174 (2003)
2. Stanford CNFET Model, http://nano.stanford.edu/models.php
3. Gojman, B., Rachlin, E., Savage, J.E.: Evaluation of Design Strategies for Stochastically Assembled Nanoarray Memories. Journal of Emerging Technologies 1(2), 73–108 (2005)
4. Heath, J.R., Ratner, M.A.: Molecular Electronics. Physics Today 56(5), 43–49 (2003)
5. Savage, J.E., Rachlin, E., DeHon, A., Lieber, C.M., Wu, Y.: Radial Addressing of Nanowires. ACM Journal of Emerging Technologies in Computing Systems 2(2), 129–154 (2006)

6. Stan, M.R., Franzon, P.D., Goldstein, S.C., Lach, J.C., Ziegler, M.M.: Molecular Electronics: From Devices and Interconnect to Circuits and Architecture. Proc. of the IEEE 91(11), 1940–1957 (2003)
7. Predictive Technology Model, http://www.eas.asu.edu/~ptm/
8. Raychowdhury, A., Roy, K.: Carbon Nanotube Electronics: Design of High Performance and Low Power Digital Circuits. IEEE Trans. on Circuits and Systems - I: Fundamental Theory and Applications 54(11), 2391–1401 (2007)
9. Williams, R.S., Kuekes, P.J.: Demultiplexer for a Molecular Wire Crossbar Network. US Patent Number 6,256,767 (2001)

A SWNT-Based Sensor for Detecting Human Blood Alcohol Concentration

H. Leng and Y. Lin*

Department of Mechanical and Industrial Engineering , Northeastern University
Boston, MA 02115 USA
yilin@coe.neu.edu

Abstract. Alcohol intake may impair human abilities, degrade human performance, and result in serious diseases. Alcohol sensors are needed to manage the risk and effect of alcohol use to human health and performance. This paper was focused on the theoretical models and design of carbon nanotube based alcohol sensors. The experiments verified that single-walled carbon nanotubes can be used to detect alcohol vapor, and need metal pads to achieve higher sensitivity.

Keywords: Sensor, blood alcohol concentration, carbon nanotube, human-machine system, driver-vehicle system.

1 Introduction and Motivations

Alcoholic beverages are popular in modern society. However, alcohol intake impairs human abilities and degrades human performance [1]. Excessive consumption of alcoholic beverages may result in serious diseases [2]. In order to manage the risk and effect of alcohol use to human health and performance, it is worthy to monitor human blood alcohol concentration (BAC). A widely acceptable method is measuring the alcohol concentration of human exhalation.

Most technologies of measuring alcohol concentration can be classified into three methods, *(1) metal oxide based methods* in which the sensing element is metal oxides such as SnO_2 [3], *(2) optical methods* in which the absorption bands of alcohol are used [4], and *(3) carbon nanotubes (CNT) based methods* in which the resistance of CNTs changes with the ambient alcohol concentration [5]. Compared to the other two methods based alcohol sensors, CNT based alcohol sensors have the potential to achieve ultra-high sensitivity, quick response, large measurement range, compact size, and low energy consumption. These features are essential to monitor human BAC for human health and performance.

This paper is focused on developing a CNT based alcohol sensor which can be used to monitor the alcohol concentration of human exhalation, and then detect human blood alcohol concentration (BAC).

M. Cheng (Ed.): NanoNet 2008, LNICST 3, pp. 69–73, 2009.

2 Theoretical Models of CNT Based Alcohol Sensors

Theoretically, there are three models to use CNTs to measure the alcohol concentration of alcohol vapor: *(1) Pure-CNT model* is to monitor the resistance of pure CNTs. Pure-CNT based alcohol sensors can be recovered using dry nitrogen gas or by heating [5], [6]; *(2) Special-CNT model* is to monitor the resistance of the CNTs that are deposited special chemical molecules (e.g. COOH). Special-CNT based alcohol sensors are recovered by heating [7]; *(3) Mixed-CNT model* is to monitor the resistance of the CNTs that are combined with enzymes. The enzymes are used to convert the alcohol into other substances which change the resistance of CNTs [8].

Pure-CNT based alcohol sensors are the simplest. Special-CNT based alcohol sensors have higher sensitivity and better gas selection than pure-CNT based alcohol sensors. Mixed-CNT model based alcohol sensors which have good gas selection require severe operating conditions. Hence, special-CNT model based alcohol sensors are more suitable for monitoring human BAC for human health and performance.

3 Design of a CNT Based Alcohol Sensor

A CNT based alcohol sensor consists of at least five parts, a substrate, a CNT sensing element, metal pads, bonding wires, and heating wires (Fig. 1). The substrate is a piece of insulating material with a central opening. The CNT sensing element is placed in the central opening, and is bonded to the pads ① and ② using bonding wires. The heating wires cover the sensing element, and build a series circuit between the pads ③ and ④. The sensing element is heated and recovered by applying a voltage between the pads ③ and ④. The recovery time from heating start to full recovery of resistance hinders the sensor to quickly and continuously respond. A feasible solution is integrating multiple sensing elements (e.g. two), a controller, a timer, and a heater into one alcohol sensor (Fig. 2). The controller, following the timer signal, can connect the signal terminals with one sensing element and recover the other sensing elements using the heater. The complexity and expense of alcohol sensor should be increased.

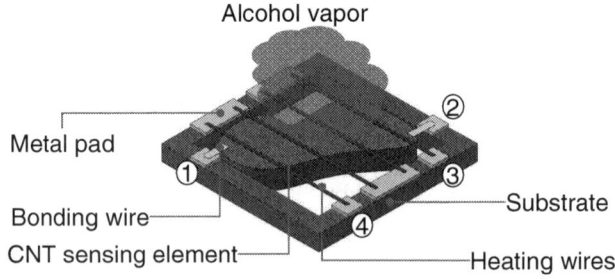

Fig. 1. A CNT based alcohol sensor

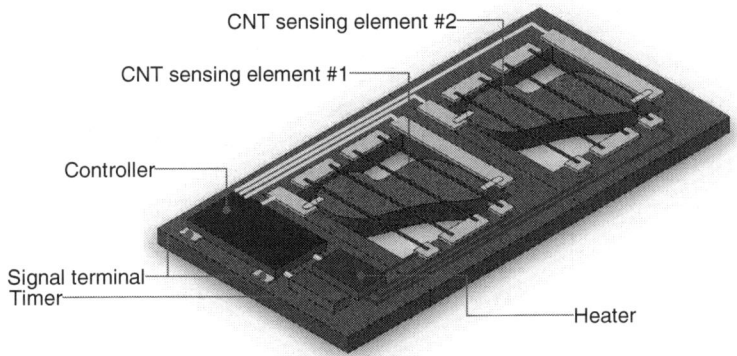

Fig. 2. A CNT based alcohol sensor combining two sensing elements

4 Experiments of CNT Sensing Elements

The functional part of the CNT based alcohol sensor in Fig.2 is still the CNT sensing element. However, it seems that researchers have not compared the performance of different CNT sensing elements in measuring alcohol vapor. Thus, this study designed two experiment groups, #1 and #2, to do the comparison.

4.1 Experiment Group #1

Experiment group #1 was carried out to verify that the resistance of CNT sensing element changes with two ambient alcohol concentrations, no alcohol and having alcohol. A multi-walled carbon nanotube (MWNT) sensing element without pads and a single-walled carbon nanotube (SWNT) sensing element without pads were prepared for the experiment group #1 using the methods of [9-10] (Fig. 3).

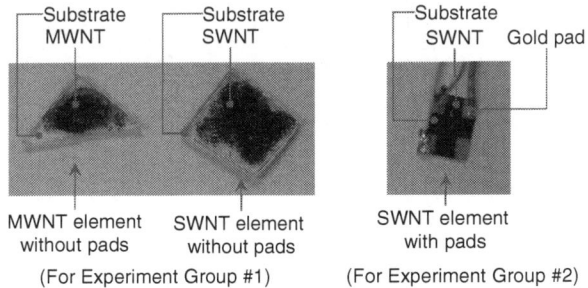

Fig. 3. CNT sensing elements for experiment group #1 and #2

Experiment apparatus and procedures. Experiment group #1 utilized ethyl alcohol (70% by volume), a precision ohmmeter, and a plastic cup with lid. The precision ohmmeter was used to measure the resistance of sensing elements. The plastic cup was utilized to establish a close space full of alcohol vapor.

When a sensing element was tested, it was fixed in the plastic cup whose lid opens, and then was connected with the ohmmeter using alligator clips. The resistance of the sensing element was recorded when the reading was stable. After placing an alcohol cotton swab in the plastic cup and closing its lid, the resistance of sensing element was recorded again when the reading was stable. Then, the resistance of the sensing element was fully recovered by removing the lid of plastic cup and the alcoholic swab. Totally, both measurements were repeated 10 times for each sensing element.

Experiment result. Experiment group #1 produced 40 records. The records of each sensing element before adding alcohol were compared with its records after adding alcohol using the analysis of variance (ANOVA) (Table 1). The result verified that the resistance of the SWNT sensing element without pads is sensitive to the ambient alcohol concentration ($\alpha=0.05$, P=0.0025). The resistance of the MWNT sensing element without pads does not significantly change with the ambient alcohol concentration ($\alpha=0.05$, P=0.7700). It is feasible to use SWNT sensing elements to measure an alcohol vapor.

Table 1. ANOVA ($\alpha=0.05$) result of experiment groups #1 and #2

Experiment Group	Sample	Source of Variation	F	P-value
#1	MWNT element without pads	no Alcohol	0.088027	0.770096
	SWNT element without pads	VS	12.36055	0.002469
#2	SWNT element with pads	having alcohol	37.47929	8.79E-06

4.2 Experiment Group #2

Experiment group #2 was performed to verify that a CNT sensing element with pads can produce more significant resistive change when the ambient alcohol concentration changes between two levels, no alcohol and having alcohol. Considering that SWNT sensing elements respond to alcohol vapor, this study prepared a SWNT sensing element with metal pads for experiment group #2 using the methods of [9-10] (Fig. 3). This sensing element was tested using the same apparatus and procedures of experiment group #1.

Experiment group #2 produced 20 records of which the ANOVA result is showed in Table 1. The P-value of the SWNT sensing element without pads ($\alpha=0.05$, P=0.0025) is much larger than the P-value of the SWNT sensing element with pads ($\alpha=0.05$, P=8.79E-06). It implies that the SWNT sensing element with pads can achieve higher sensitivity in measuring an alcohol vapor.

5 Conclusions and Future Work

When human blood alcohol concentration is measured to decrease the effect of alcohol use to human health and performance, the severe technical requirements can be satisfied using carbon nanotube based alcohol sensors. Among their three theoretical models, the special carbon nanotube based alcohol sensor is more suitable for this measurement. The recovery time of carbon nanotube based alcohol sensor can be eliminated by integrating multiple sensing elements into one sensor. The experiments verified that single-walled carbon nanotubes can be used to detect an alcohol vapor, whereas multi-walled carbon nanotubes can not. The single-walled carbon nanotubes with metal pads have higher sensitivity than those without pads.

The future work is to produce a SWNT based alcohol sensor involving two sensing elements and test its performance.

Acknowledgments. The carbon nanotube samples used in this study was provided by Dr. Yung Joon Jung and Laila Jaberansari.

This project is supported by National Science Foundation (NSF) the Sensor Innovation and Systems program (Award Number 0825864). The authors would like to thank Dr. Shih-Chi Liu for his kind support to this research.

References

1. Kim, H., Yang, C., Lee, B., Yang, Y., Hong, S.: Alcohol effects on navigational ability using ship handling simulator. International Journal of Industrial Ergonomics 37(9-10), 733–743 (2007)
2. National Institute on Alcohol abuse and Alcoholism (NIAAA), 2006, Report to the extramural advisory board (2006),
   `http://pubs.niaaa.nih.gov/publications%5CDEPRStrategicPlan%`
   `5CBriefingBook2.htm#D._CHRONIC_DISEASES`
3. Promsong, L., Sriyudthsak, M.: Thin tin-oxide film alcohol-gas sensor. Sensors and Actuators B (Chemical) B25(1-3), 504–506 (1995)
4. Azzazy, M., Chau, T., Wu, M., Tanbun-Ek, T.: Remote sensor to detect alcohol impaired drivers. IEEE Lasers and Electro-Optics Society 1995 Annual Meeting. In: 8th Annual Meeting. Conference Proceedings, vol. 2, pp. 320–321 (1995) (Cat. No.95CH35739)
5. Someya, T., Small, J., Kim, P., Nuckolls, C., Yardley, J.: Alcohol Vapor Sensors Based on Single-Walled Carbon Nanotube Field Effect Transistors. Nano letters 3(7), 877–881 (2003)
6. Kong, J., Franklin, N.R., Zhou, C., Chapline, M.G., Peng, S., Cho, K., Dai, H.: Nanotube molecular wires as chemical sensors. Science 287(5453), 622–625 (2000)
7. Sin, M., Chow, G., Wong, G., Li, W., Leong, P., Wong, K.: Ultralow-power alcohol vapor sensors using chemically functionalized multiwalled carbon nanotubes. IEEE Transactions on Nanotechnology 6(5), 571–577 (2007)
8. Tsai, Y., Huang, J., Chiu, C.: Amperometric ethanol biosensor based on poly(vinyl alcohol)-multiwalled carbon nanotube-alcohol dehydrogenase biocomposite. Biosensors & Bioelectronics 22(12), 3051–3056 (2007)
9. Jung, Y.: Controlled Synthesis of Carbon Nanotubes using Chemical Vapor Deposition Methods. In: Nanomanufacturing Handbook, ch. 4. CRC, Boca Raton (2006)
10. Xiong, X., Jaberansari, L., Hahm, M., Busnaina, A., Jung, Y.: Building Highly Organized Snigle-Walled Carbon Nanotube Networks Using Template Guided Fluidic Assembly. Small 3(12), 2006–2010 (2007)

A Dual-Mode Hybrid ARQ Scheme for Energy Efficient On-Chip Interconnects

Bo Fu and Paul Ampadu

Department of Electrical and Computer Engineering
University of Rochester, Rochester, 14627, USA
{bofu, ampadu}@ece.rochester.edu

Abstract. In this paper, we propose a dual-mode hybrid ARQ scheme for energy efficient on-chip communication, where the type of coding scheme can be dynamically selected based on different noise environments and reliability requirements. In order to reduce codec area overhead, a hardware sharing design method is implemented, resulting in only a minor increase in area costs compared to a single-mode system. For a given reliability requirement, the proposed error control scheme yields up to 35% energy improvement compared to previous solutions and up to 18% energy improvement compared to worst-case noise design method.

Keywords: Adaptive error control, on-chip interconnects, hybrid ARQ, interleaving.

1 Introduction

On-chip interconnect errors, exacerbated by very-deep submicron (VDSM) technology, can be caused by supply voltage fluctuation, crosstalk, process variation, radiation and electromagnetic interference [1]. Error control schemes, such as automatic repeat request (ARQ), forward error correction (FEC) and hybrid ARQ (HARQ), are widely used to improve reliability of on-chip interconnects in VDSM technology [1]-[6]. Each error control scheme has different area, power, throughput, and error correction capability trade-offs. Selection of the proper scheme can be a design-time decision based on quality-of-service (QoS) requirements [2,3]; however, noise conditions vary with different environmental factors (e.g. temperature) and operational conditions (e.g. supply voltage), and designing for the worst case can waste energy [5]. To achieve energy efficiency, an error control scheme is needed which can intelligently provide appropriate error control capability based on noise conditions or system requirements [5,6].

In this paper, we focus on HARQ schemes, which combine FEC and ARQ to achieve a good balance of reliability and energy consumption [1,2]. The proposed dual-mode HARQ scheme can be dynamically configured in different noise environments. Further, in order to reduce the codec area overhead, a hardware sharing method is introduced. The proposed dual-mode HARQ method is presented in Section 2. In Section 3, the design is evaluated in terms of reliability, area and energy consumption. Conclusions are presented in Section 4.

M. Cheng (Ed.): NanoNet 2008, LNICST 3, pp. 74–79, 2009.

2 Proposed Error Control Scheme

In [1,2], single-error correcting double-error detecting (SEC-DED) codes (e.g. extended Hamming) are used to perform HARQ. As technology scales, spatial burst errors, where several adjacent bus lines are erroneous, become more common because of crosstalk effects [1]. In order to improve error resilience against burst errors, a wide bus can be split into smaller groups and encoded separately, and the outputs of these small groups can be interleaved to reduce the probability of multiple errors occurring within the same group [3]. Unfortunately, separating into groups and interleaving increase link energy consumption because of the large wire requirements.

In this paper, we propose a dual-mode HARQ scheme, which combines a traditional HARQ scheme using SEC-DED codes with interleaving. The proposed method works in two modes–(a) directly using the traditional HARQ method, or (b) separating the input message into four groups and encoding each group with a SEC-DED code, then interleaving the outputs of each group. In order to reduce the area overhead, a hardware sharing method is introduced. Fig. 1 shows the block diagram of the proposed dual-mode HARQ scheme. The input information is split into four identical groups. Each group is encoded with extended Hamming codes. The parity check bits of each group can be directly sent to the interleaver (mode b) or combined to generate the parity check bits of another SEC-DED code which uses the whole message as an input (mode a). MUXs are used to select the appropriate parity check bits for a mode based on control signals, which can be generated using the method in [5]. Fig. 2 shows an example of the interleaving relationship for a 16-bit input message.

The following example demonstrates the proposed dual-mode HARQ design. Consider a 16-bit input message, which is separated into four groups. Each group is encoded using an extended Hamming code H(8,4) with the generator matrix in Eq. (1). The parity check bits of each group can be combined to generate parity check bits of an extended Hamming code H(22,16) with the generator matrix in Eq. (3), where $P_{16\times6}^T$ is the transpose of the parity matrix. The parity matrix

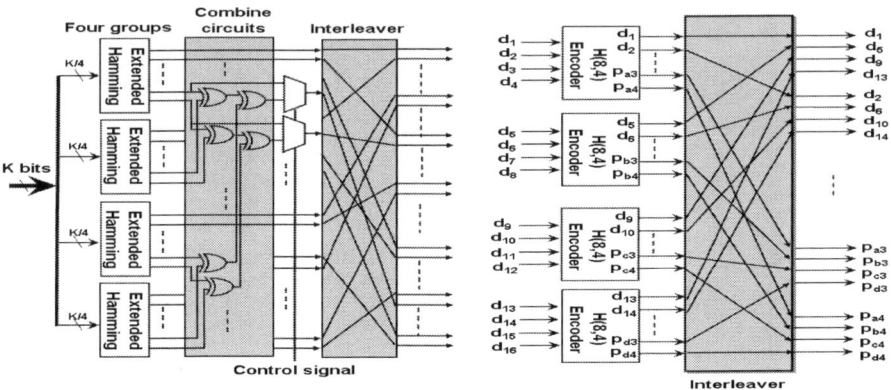

Fig. 1. Proposed dual-mode HARQ scheme   **Fig. 2.** Interleaving for 16-bit information

$$G_1 = [\, I_{4\times4} \mid P_{4\times4} \,] = \begin{bmatrix} 1\,0\,0\,0 & 1\,1\,0\,1 \\ 0\,1\,0\,0 & 1\,1\,1\,0 \\ 0\,0\,1\,0 & 1\,0\,1\,1 \\ 0\,0\,0\,1 & 0\,1\,1\,1 \end{bmatrix} \quad (1)$$

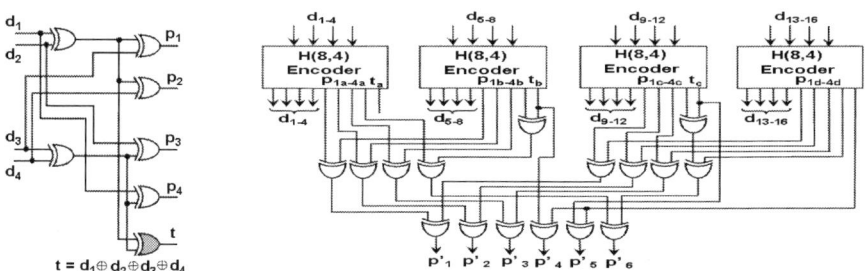

$$P_{4\times4}^{T} = \begin{bmatrix} 1\,1\,1\,0 \\ 1\,1\,0\,1 \\ 0\,1\,1\,1 \\ 1\,0\,1\,1 \end{bmatrix} \begin{array}{l} \}\,M_1 \\ \\ \}\,M_2 \end{array} \quad (2)$$

$$G_2 = [\, I_{16\times16} \mid P_{16\times6} \,]$$

$$P_{16\times6}^{T} = \begin{bmatrix} 1\,1\,1\,0 & 1\,1\,1\,0 & 1\,1\,1\,0 & 1\,1\,1\,0 \\ 1\,1\,0\,1 & 1\,1\,0\,1 & 1\,1\,0\,1 & 1\,1\,0\,1 \\ 0\,1\,1\,1 & 0\,1\,1\,1 & 0\,1\,1\,1 & 0\,1\,1\,1 \\ 0\,0\,0\,0 & 1\,1\,1\,1 & 0\,0\,0\,0 & 1\,1\,1\,1 \\ 0\,0\,0\,0 & 0\,0\,0\,0 & 1\,1\,1\,1 & 1\,1\,1\,1 \\ 1\,0\,1\,1 & 0\,1\,0\,0 & 0\,1\,0\,0 & 1\,0\,1\,1 \end{bmatrix} = \begin{bmatrix} M_1 & M_1 & M_1 & M_1 \\ 0\,0\,0\,0 & 1\,1\,1\,1 & 0\,0\,0\,0 & 1\,1\,1\,1 \\ 0\,0\,0\,0 & 0\,0\,0\,0 & 1\,1\,1\,1 & 1\,1\,1\,1 \\ M_2 & \sim M_2 & \sim M_2 & M_2 \end{bmatrix} \quad (3)$$

$$\underbrace{}_{\text{group a}} \quad \underbrace{}_{\text{group b}} \quad \underbrace{}_{\text{group c}} \quad \underbrace{}_{\text{group d}}$$

of H(22,16) is constructed as follows: the first three rows in $P_{16\times6}^{T}$ of H(22,16) are duplications of the first three rows in $P_{4\times4}^{T}$ of H(8,4) (shown in Eq. (2)); the fourth and fifth rows in $P_{16\times6}^{T}$ are either four zeros or four ones; the last row in $P_{16\times6}^{T}$ is either the duplication of the last row in $P_{4\times4}^{T}$ or its inverse. The hardware implementation of the H(8,4) and H(22,16) encoder is shown in Fig. 3 and Fig. 4. A pattern of four ones in the fourth or fifth row requires the XOR of those four data bits, implemented by adding one extra XOR gate in each H(8,4) encoder, shown as the shaded XOR gate in Fig. 3.

Fig. 3. H(8,4) encoder   **Fig. 4.** Implementation of H(22,16) encoder

3   Results and Analysis

In this section, the performance of the proposed dual-mode HARQ is evaluated. A 64-bit input message is used, which can be encoded using one extended Hamming code H(72,64) or four extended Hamming codes H(22,16). A 45 nm Predictive Technology Model (PTM) [7] is used and the link length is 3 mm with feature sizes from [8]. The clock frequency is 1 GHz. The residual flit error rate, which is the probability of error after decoding, is used to evaluate the reliability of the proposed method. The error probability of a single wire can be modeled by below [9],

$$\varepsilon = Q\left(\frac{V_{DD}}{2\sigma_N}\right) = \int_{\frac{V_{DD}}{2\sigma_N}}^{\infty} \frac{1}{\sqrt{2\pi}} e^{-y^2/2}\, dy \quad (4)$$

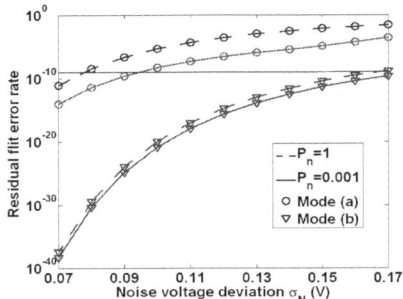

Fig. 5. Residual flit error rate

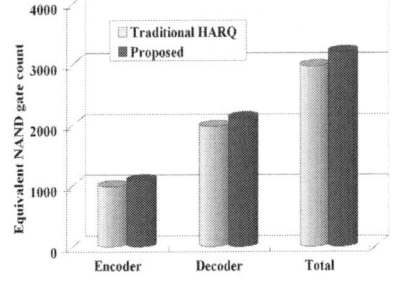

Fig. 6. Area comparison

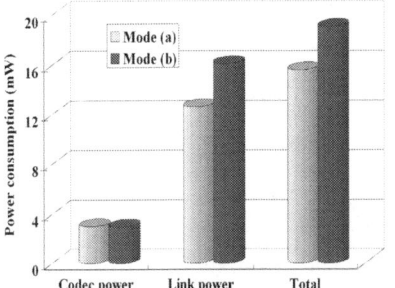

Fig. 7. Power consumption

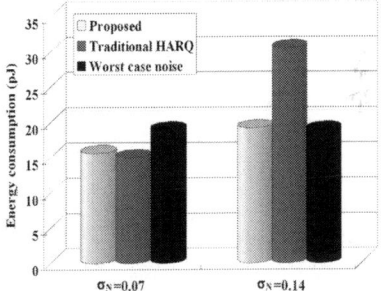

Fig. 8. Energy comparison

where V_{DD} is the supply voltage and σ_N is the standard deviation of the noise voltage, which is assumed to be a normal distribution. This model can be extended to describe multiple adjacent errors by assuming that a fault affects its neighboring wires with a certain probability P_n. Fig. 5 shows the residual flit error rate of the proposed dual-mode HARQ scheme working at different modes for different P_n values. The supply voltage is 1 V. The results show that mode (b) (separating into groups) achieves a significant improvement in residual flit error rate compared to mode (a) (traditional HARQ) when spatial burst errors are considered. For a 10^{-9} residual flit error rate requirement, mode (a) satisfies the requirement for $\sigma_N < 0.1$. Mode (b) satisfied the requirement up to $\sigma_N \cong 0.17$.

Fig. 6 compares the area of the proposed dual-mode scheme to a traditional HARQ with extended Hamming code H(72,64) in terms of equivalent NAND gate count. The results show that the codec area of the proposed method increases by about 10% compared to the traditional HARQ scheme. The codec power also increases about 10% compared to the traditional HARQ scheme. Fig. 7 shows the codec power and link power of the proposed method in each mode. The results show that link power dominates the total power consumption. Mode (b) consumes 18% more power compared to mode (a) because of the larger number of link wires used.

Fig. 8 evaluates the energy consumption of the proposed scheme for a residual flit error rate requirement of 10^{-9}. The results are compared to a traditional HARQ scheme using H(72,64) as well as worst-case noise design, in which four groups are always used. Two noise environments are considered. For the low noise environment (σ_N=0.07), the proposed dual-mode HARQ method works in mode (a) and consumes similar energy to the traditional HARQ. Compared to worst-case noise design, the proposed method achieves about 18% reduction in energy consumption. In the high noise environment (σ_N=0.14), the proposed dual-mode HARQ scheme switches to mode (b). In order to meet this reliability requirement, traditional HARQ needs a higher link swing voltage [1,10], which greatly increases the link energy. In the high noise environment, the proposed method consumes similar energy to worst-case noise design. Compared to previous solutions to increase the link swing voltage of the tradition HARQ scheme, the proposed dual-mode HARQ scheme can achieve 35% improvement in energy consumption.

4   Conclusion

In this paper, a dual-mode HARQ scheme is proposed for energy efficient on-chip communication. The efficient combination of a traditional HARQ scheme with interleaving shows a good balance between reliability and energy efficiency when burst errors are considered. In order to reduce codec area overhead, a hardware sharing design method is implemented and leads to a minor increase in area costs.

The type of error correction code can be dynamically selected in the proposed dual-mode HARQ scheme based on different noise environments. For a given system reliability requirement, the proposed error control scheme yields up to 35% energy improvement compared to previous solutions or up to 18% energy improvement compared to designing for worst-case noise.

References

1. Bertozzi, D., Benini, L., De Micheli, G.: Error control schemes for on-chip communication links: the energy-reliability tradeoff. IEEE Trans. Computer-Aided Design Integr. Circuits Syst. 24, 818–831 (2005)
2. Murali, S., et al.: Analysis of error recovery schemes for networks-on-chips. IEEE Des. Test Comput. 22, 434–442 (2005)
3. Zimmer, H., Jantsch, A.: A fault model notation and error-control scheme for switch-to-switch buses in a network-on-chip. In: International Conference on Hardware/Software Codesign and System Synthesis, pp. 188–193 (2003)
4. Sridhara, S., Shanbhag, N.R.: Coding for system-on-chip networks: a unified framework. IEEE Trans. Very Large Scale Integr (VLSI) Syst. 13, 655–667 (2005)
5. Li, L., Vijaykrishnan, N., Kandemir, M., Irwin, M.J.: Adaptive error protection for energy efficiency. In: IEEE/ACM International Conference on Computer-Aided Design, pp. 2–7 (2003)
6. Rossi, D., Angelini, P., Metram, C.: Configurable error control scheme for NoC signal integrity. In: International On Line Testing Symposium, pp. 43–48 (2007)

7. Arizona State Univ., Predictive Technology Model,
   `http://www.eas.asu.edu/~ptm/`
8. Xu, S., Benito, I., Burleson, W.: Thermal impacts on NoC interconnects. In: IEEE
   International Symposium on Networks-on-Chip, pp. 220 (2007)
9. Hegde, R., Shanbhag, N.R.: Towards achieving energy-efficiency in presence of deep
   submicron noise. IEEE Trans. Very Large Scale Integr (VLSI) Syst. 8, 379–391
   (2000)
10. Worm, F., et al.: A robust self-calibrating transmission scheme for on-chip net-
    works. IEEE Trans. Very Large Scale Integr. (VLSI) Syst. 13, 126–139 (2005)

Using Randomly Assembled Networks for Computation*

Andrey Zykov and Gustavo de Veciana

Department of Electrical & Computer Engineering
The University of Texas at Austin

Abstract. This paper makes the case for perturbation-based computational model as a promising choice for implementing next generation
ubiquitous information applications on emerging nanotechnologies. Our
argument centers on its suitability for technologies with low manufacturing precision, high defect densities and performance uncertainty. This
paper discusses some of the possible advantages and pitfalls of this approach, and associated novel design principles.

Keywords: perturbation, computation model, embedded systems.

1 Introduction

Advances in the synthesis and self-assembly of nanoelectronic devices suggest
that the ability to manufacture dense nanofabrics is on the near horizon, see e.g.,
[1,2]. Yet, effective ways of utilizing emerging nanoelectronic technologies still
elude us. The tremendous increase in device density afforded by nanotechnologies
is expected to be accompanied by substantial increases in defect densities, performance variability, and susceptibility to single event upsets caused by cosmic
radiation (energetic neutrons) and alpha particles. System-level design adher-
ing to current computational models may thus soon reach fundamental scaling
limits, where the increased densities are countered by overheads associated with
achieving defect- and fault-tolerant designs that are robust to performance variability. Thus, it is critical to consider and explore alternative computational
models that can operate under such difficult conditions.

In this paper, we investigate a promising new class of computational model,
called perturbation-based and show its potential to synergistically address the
two sides of the complex system design equation: technology and applications.
On one hand we argue it suitability in overcomming the reliability challenges
associated with emerging nanoelectronics. On the other hand we focus on meeting the needs and leveraging the characteristics of emerging, but soon to be
ubiquitous, soft real-time processing and control applications.

---

* This work is supported by the Gigascale Systems Research Center under the 'Alternative' Theme.

2   Principles of Perturbation-Based Computing

Perturbation-based computational models are ideal for implementing complex non-linear filters (operators) associated with real-time information processing. The key idea is to perform a non-linear projection of the input stream into a high dimensional space using a complex dynamical system. If the pool of dynamics capturing information about current and past stimuli is sufficiently rich, *any* desired non-linear filtering task's output(s) can be derived, or 'composed' from it. Below we develop this basic idea in a more rigorous manner. Note this computational models was recently independently discovered by two research groups [3,4]. Yet, their work was driven by research pursuits and objectives quite different from those in this paper.

Perturbation-Based Machines. Fig. 1 symbolically depicts a perturbation-based machine M. As can be seen, it maps an input function $u(\cdot)$ to an output function $y(\cdot)$, relying on two key components: a high dimensional dynamical system, implementing the machine's *computational core* D^M, and an *output stage* f^M. The key premise underlying perturbation-based computing is that, by using computational cores realized by sufficiently complex, even random, dynamical systems, one can essentially project inputs over a sufficiently large family of basis operators for any given set applications and desired approximation level [3]. A machine's D^M is thus a dynamical system realizing a very large pool of candidate operators, while the above mentioned D^M denotes a specific set of basis operators required for a given approximation. As such, the *same* computational core D^M can be used in realizing various tasks. The output stage is the *task dependent* part of this machine, playing the role of both selecting and composing the 'relevant' basis operators through a memoryless function. As shown in Fig. 1, the computational core D^M generates an internal state $x^M(t)$, corresponding to a causal response to the input u. This is a non-linear projection of the input stream on a high dimensional space, generated by exciting the dynamical system associated with D^M. Note that *no stable internal states* are required in the computational core, it suffices to generate a sufficiently rich pool of transient dynamics. The output stage f^M maps the internal transient state to the desired output. Note that the precise internal dynamics of the core network need not have a specific form, as long as the core projects the input signals on a sufficiently rich set of basis functions.

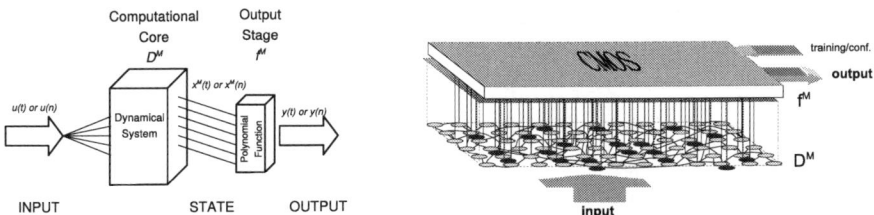

Fig. 1. On the left a Perturbation-based machine. On the right a proposed hybrid eNano-CMOS configurable platform for perturbation-based computing.

Universal Approximative Power and target applications. Based the work in [5], [3] established that perturbation-based machines have universal computing power – that is, machines operating 'natively' under this computational model can approximate *arbitrarily closely* any time invariant fading memory operator. Still, although this result tells us that the number of basis operators required by any such approximation is *finite*, it says nothing about how many such operators may be required in each case. If very high precision is required, the number of operators may be very high for certain tasks. The proposed computational model is inherently based on approximation. Therefore it is not expected to operate without errors, with perhaps the exception of approximation of very simple functions/operators. Although for some tasks this will be unacceptable, for others this presents an opportunity to tradeoff error rate against other costs, e.g., manufacturing cost, power consumption etc. An example of such tasks would be those involving real-time searches for opportunities, e.g., block matches across frames in video compression. If we miss an opportunity this will not cause algorithm failure, instead it results in a temporarily lower compression rate. Another class of applications involves systems with feedback, where such small/rare errors can be compensated subsequently through feedback overall having a negligible effect. More generally the aim is not to achieve high precision, but rather simplicity and universality. Blocks having moderate reliability can be bootstrapped to construct more complex and/or reliable operators, e.g., through averaging or other forms of aggregation [6] or specific mechanisms akin to the way complex logic functions are constructed from elementary logic gates (e.g. NAND, NOR, etc).

3   Design Principles for Perturbation Based Computing

We begin this section by first proposing a hybrid eNano-CMOS platform as a representative realization of perturbation-based machines. The machine's computational core is implemented on an emerging nanoelectronic fabric while CMOS is used to implement the simple (e.g., linear) read out function at the output stage and support the machine configuration/training process. Fig. 1 shows an abstract view of such a platform, with the key basis operators in the pool highlighted in bold. Clearly, this platform can directly leverage the formidable densities achieved by nanotechnologies to create computational cores of essentially arbitrary size. At the same time the more reliable CMOS layer allows to reliably configure (train) the output readouts to properly approximate the desired function. We discuss five underlying design principles next and refer the reader to [7] for experimental results supporting them:

1. Defect-Tolerance, Randomly Assembled Cores, and Training. First we need only ensure that the core and readout connectivity are sufficiently 'rich' to achieve the desired approximation after training. The designer needs only control the size and statistics of the core network without precisely specifying its topology. Manufacturing defects and heterogeneity in the network become part of its intrinsic randomness. This flexibility comes at the cost of performing a configuration/training step for each chip – which is indeed a costly requirement.

Yet it can be viewed as 'similar' to the overheads associated with typical defect tolerance approaches. Indeed the typical requirements in the latter are to detect, i.e., map out, defects for each chip and then re-synthesize the function to avoid defects. Defect mapping is typically done using test patterns that are either obtained/generated off chip or stored on chip. Re-synthesis involves reprogramming the function around the defects on the chip. In our case rather than defect mapping and re-synthesis steps we require a training step. Such training will involve access to input-output pairs that can also be provided either off-chip or on-chip. A comparison of the cost of mapping an re-synthesis vs training is premature. Another potential disadvantage is possibility of excessive power consumption by random core in contrast to precisely controlled core. However reduced control requirements also open opportunities to make these random cores very cheap especially in terms of power consumption.

2. Fault-Tolerance Through Unstructured Redundancy. Second, fault tolerance can also be partially achieved by appropriately defining the statistics and size of the core. Intuitively, even if randomly assembled, a large dynamical network should incorporate sufficient redundancy to allow the readout layer to average out internal noise/soft errors. We refer to this as unstructured redundancy in the core, as it need not be explicitly designed, e.g., as would be the case with triple modular redundancy. Instead the designer need only decide on a sufficiently large core to address soft faults and/or internal performance variability, e.g., due to coupling etc. An obvious advantage of this approach is reduction in design cost in comparison to structured redundancy. A disadvantage this comes at a cost, bigger cores will consume more power.

3. Complete Core Sharing. Note that aside from general considerations on size and network statistics detailed core characteristics are task independent. Thus several different tasks that share the same input can in principle share its projection on the same core. For example word recognition and speaker identification tasks for the same speech input could share the same core. Such complete and parallel sharing of resources has the potential to substantially reduce overall system cost in terms of both area and power. Note however that the readout layer can not be easily shared across tasks, which may lead to a scalability problem if a core is to be shared among a large number of tasks.

4. Weakly Interconnected Networks and Spatial Decomposition. A potential problem with scaling to large cores is a scalability problem if random interconnections among nodes are necessary. We propose a fourth design principle towards overcoming this problem. The idea is that to introduce some hierarchy by only weakly interconnecting smaller cores. This allows one to control the interconnect costs as the size of the cores increase. Moreover this seems a natural way to randomly assemble cores, i.e., one where the primary form of connectivity is local. More generally one can imagine designs that leverage a large number of relatively small cores which serve as building blocks to create bigger cores as needed.

5. Nearly Decomposable Core Dynamics. The last design principle we propose relates to decomposition in terms of temporal dynamics. The idea is that some applications are driven by (possibly coupled) dynamics at different time scales, which a designer might recognize and incorporate into his core design. For example a core design might include weakly interconnected cores operating at different speeds. One can imagine, creating cores with different response times to input signals, through some form of doping and/or processing. For applications exhibiting dependencies on multiple time scales such decompositions are very effective at reducing complexity. Furthermore purposefully combining fast and slow cores may present further advantages towards reducing power consumption. Note that the principle here is not perform careful core design, but simply define some large scale characteristics for connectivity and dynamics of its constituent subnetworks.

4   Conclusion

This paper explores an alternative computational model for "nanocomputing". We have pointed out a variety of interesting characteristics, such as the possibility of using randomly assembled nano network cores for computation. This approach is consistent with circumventing what appears to be an intrinsic difficulty and thus costly requirement of precisely controlling the manufacturing of nano systems. So the question is to what degree can one give up on precise control over manufacturing and still devise useful computation systems? In this paper we presented one possible approach but there may be others (perhaps more general) models.

References

1. Bourianoff, G.: The future of nanocomputing. Computer Magazine, 44–49 (August 2003)
2. Sematech. International technology roadmap for semiconductors - 2004 update on emerging research devices (2004),
   http://www.itrs.net/Common/2004Update/2004Update.htm
3. Maass, W., Natschlager, T., Markram, H.: Real-time computing without stable states: A new framework for neural computation based on perturbations. Neural Computation 14(11), 2531–2560 (2002)
4. Jaeger, H.: The echo state approach to analysing and training recurrent neural networks. GMD-Report 148, German Nat. Res. Inst. Comp. Sci. (2001)
5. Boyd, S., Chua, L.: Fading memory and the problem of approximating nonlinear operators with volterra series. IEEE Trans. Circ. Sys. 32, 1150–1161 (1985)
6. Varatkar, G.V., Narayanan, S., Shanbhag, N., Jones, D.L.: Sensor network-on-chip. In: International Symposium on SOC (November 2007)
7. Zykov, A., Jacome, M., de Veciana, G.: Perturbation-based computing for next-generation embedded IT targeted at emerging nanoelectronics (submitted) (2008)

A Biochemically-Engineered
Molecular Communication System
(Invited Paper)

Satoshi Hiyama[1,*], Yuki Moritani[1], and Tatsuya Suda[1,2]

[1] Research Laboratories, NTT DOCOMO, Inc.
3-6 Hikarinooka, Yokosuka-shi, Kanagawa 239-8536 Japan
[2] Information and Computer Science, University of California,
Irvine, CA 92697-3425 USA
{hiyama,moritani,suda}@nttdocomo.co.jp
suda@ics.uci.edu

Abstract. Molecular communication uses molecules (i.e., chemical signals) as an information carrier and allows biologically- and artificially-created nano- or cell-scale entities to communicate over a short distance. It is a new communication paradigm and is different from the traditional communication paradigm that uses electromagnetic waves (i.e., electronic and optical signals) as an information carrier. Key research challenges in molecular communication include design of a sender, design of a molecular propagation system, design of a receiver, design of a molecular communication interface, and mathematical modeling of molecular communication components and systems. This paper focuses on system design and experimental results of molecular communication and briefly refers to recent activities in molecular communication.

Keywords: Nanotechnology, Bioengineering, Biochemical communication system, Functional soft materials.

1 Introduction

Molecular communication [1]-[2] is inspired by the biological communication mechanisms (e.g., cell-cell communication using hormones) [3] and artificially creates a biochemically-engineered communication system in which communication processes are controllable. Molecular communication uses molecules (i.e., chemical signals) as an information carrier and allows biologically- and artificially-created nano- or cell-scale entities (e.g., cells and biohybrid devices) to communicate over a short distance. It is a new communication paradigm and is different from the traditional communication paradigm that uses electromagnetic waves (i.e., electronic and optical signals) as an information carrier.

In molecular communication, a sender encodes information onto molecules (called information molecules) and emits the information molecules to the propagation environment. A propagation system transports the emitted information molecules to a

M. Cheng (Ed.): NanoNet 2008, LNICST 3, pp. 85–94, 2009.
© ICST Institute for Computer Sciences, Social Informatics and Telecommunications Engineering 2009

receiver. The receiver, upon receiving the transported information molecules, reacts biochemically to the received information molecules (this biochemical reaction represents decoding of the information).

Molecular communication is a new and interdisciplinary research area that spans the nanotechnology, biotechnology, and communication technology, and as such, it requires research into a number of key areas. Key research challenges in molecular communication include 1) design of a sender that generates molecules, encodes information onto the generated molecules, and emits the information encoded molecules (information molecules), 2) design of a molecular propagation system that transports the emitted information molecules from a sender to a receiver, 3) design of a receiver that receives the transported information molecules and biochemically reacts to the received information molecules, 4) design of a molecular communication interface between a sender and a molecular propagation system and also between the propagation system and a receiver to allow a generic transport of information molecules independent of their biochemical/physical characteristics, 5) mathematical modeling of molecular communication components and systems. This paper focuses on system design and experimental results of molecular communication.

The rest of this paper is organized in the following manner. Section 2 presents key features and basic communication processes in molecular communication. Section 3 explains detailed system design and initial experimental results of key components in a molecular communication system. Section 4 briefly describes selected recent activities in molecular communication and concludes the paper.

2   Key Features and Basic Communication Processes

Molecular communication is a new communication paradigm and is different from the traditional communication paradigm (Table 1). Unlike the traditional communication that utilizes electromagnetic waves as an information carrier, molecular communication utilizes molecules as an information carrier. In addition, unlike in the traditional communication where encoded information such as voice, text, and video is decoded and regenerated at a receiver, in molecular communication, information molecules cause some biochemical reactions at a receiver and recreate phenomena and/or chemical status that a sender transmits. Other features of molecular communication include aqueous environmental communication, stochastic nature of communication, low energy-consumption communication, and being highly compatible with biological systems.

Although the communication speed/distance of molecular communication is slower/shorter than that of the traditional communication, molecular communication may carry information that is not feasible to carry with the traditional communication (such as biochemical status of a living organism) between the entities that the traditional communication does not apply (such as biological entities). Molecular communication has unique features that are not seen in the traditional communication and is not competitive but complementary to the traditional communication.

Figure 1 depicts an overview of a molecular communication system that includes senders, molecular communication interfaces, molecular propagation systems, and receivers.

Table 1. Comparisons of key features between the traditional communication and molecular communication

Key features	Traditional communication	Molecular communication
Information carrier	Electromagnetic waves	Molecules
Signal type	Electronic and optical signals	Chemical signals
Propagation speed	Light speed (3×10^5 km/sec)	Slow speed (a few μm/sec)
Propagation distance	Long (ranging from m to km)	Short (ranging from nm to m)
Propagation environment	Airborne and cable medium	Aqueous medium
Encoded information	Voice, text, and video	Phenomena and chemical status
Behavior of receivers	Decoding of digital info	Biochemical reaction
Communication model	Deterministic communication	Stochastic communication
Energy consumption	High	Extremely low

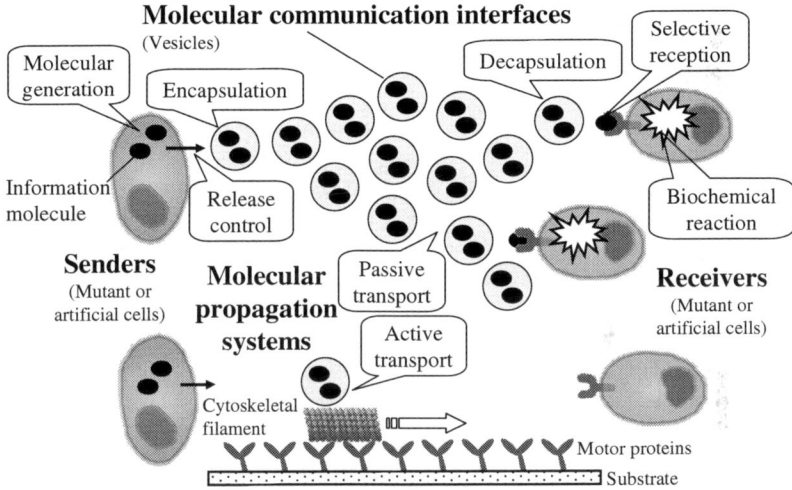

Fig. 1. An overview of a molecular communication system

A sender generates molecules, encodes information onto the generated molecules, and emits the information encoded molecules (information molecules) into a propagation environment. The sender may encode information on the type of the information molecules used or the concentration of the information molecules used. Possible approaches to create a sender include genetically modifying eukaryotic cells and artificially constructing biological devices that are capable of performing the encoding.

A molecular communication interface acts as a molecular container that encapsulates information molecules to hide the characteristics of the information molecules during the propagation from the sender to a receiver to allow a generic transport of information molecules independent of their biochemical/physical characteristics. Using a lipid bilayer vesicle [4] is a promising approach to encapsulate the information molecules. Encapsulated information molecules are decapsulated at a receiver.

A molecular propagation system passively or actively transports information molecules (or vesicles that encapsulate information molecules) from a sender to an appropriate receiver through the propagation environment. The propagation environment is aqueous solution that is typically found within and between cells. Using biological motor systems (motor proteins and cytoskeletal filaments) [5] are a promising approach to actively and directionally transport information molecules.

A receiver selectively receives transported and decapsulated information molecules, and biochemically reacts to the received information molecules. Possible approaches to create a receiver are to genetically modify eukaryotic cells and to artificially construct biological devices as to control the biochemical reaction.

3   Detailed System Design and Initial Experimental Results

This section describes detailed system design and initial experimental results of key components in a molecular communication system, and shows that our system design is feasible.

3.1   Molecular Communication Interface

A vesicle-based communication interface provides a mechanism to transport different types of information molecules in diverse propagation environments [6]. This is because the vesicle structure (i.e., a lipid bilayer membrane) provides a generic architecture that compartmentalizes and transports diverse types of information molecules independent of their biochemical/physical characteristics. The vesicle structure also protects information molecules from denaturation (e.g., molecular deformation caused by changes in temperature or pH) in the propagation environment. Key research issues in implementing the vesicle-based communication interface include how vesicles encapsulate information molecules at a sender and how vesicles decapsulate the information molecules at a receiver.

The authors of this paper have proposed a molecular communication interface that uses a vesicle embedded with gap junction proteins (Fig. 2) [7]. A gap junction is an inter-cellular communication channel formed between neighboring two cells, and it consists of two docked hemichannels (connexons) constructed from self-assembled six gap junction proteins (connexins) [8]. When a gap junction is open, molecules whose molecular masses are less than 1.5 kDa can directly propagate through the gap junction channel connecting two cells according to the molecular concentration gradient. A gap junction hemichannel is closed unless two hemichannels are docked.

In the molecular communication interface that the authors of this paper proposed, a sender stores information molecules inside itself and has gap junction hemichannels. When a vesicle with gap junction hemichannels physically contacts the sender, gap junction channels are formed between the sender and the vesicle, and the information molecules are transferred from the sender to the vesicle according to the molecular concentration gradient. When the vesicle detaches from the sender spontaneously, the gap junction hemichannels at the sender and at the vesicle close, and the information molecules transferred from the sender to the vesicle are encapsulated in the vesicle. Encapsulation of information molecules in a vesicle allows a molecular propagation system to transport the information molecules from the sender to a receiver independent of their biochemical/physical characteristics. A receiver also has gap junction

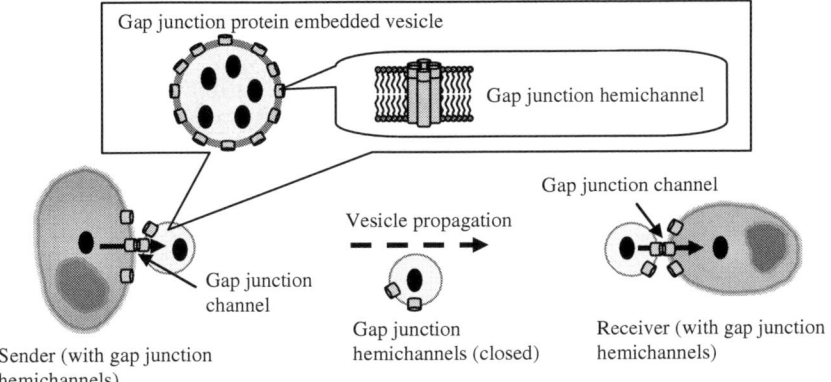

Fig. 2. A schematic diagram of a molecular communication interface using a vesicle embedded with gap junction proteins

hemichannels, and when the transported vesicle physically contacts the receiver, a gap junction channel is formed between the vesicle and the receiver, and the information molecules in the vesicle are transferred into the receiver according to the molecular concentration gradient.

In order to investigate the feasibility of the designed communication interface, the authors of this paper created connexin-43 (one of the gap junction proteins) embedded vesicles [7]. Microscopic observations confirmed that calceins (hydrophilic dyes used as model information molecules) were transferred between connexin-43 embedded vesicles and the transferred calceins were encapsulated into the vesicles [9]. This result indicates that the created connexin-43 embedded vesicle (a molecular communication interface) may encapsulate information molecules and receive/ transfer information molecules from/into a sender/receiver through gap junctions.

3.2 Molecular Propagation System

In eukaryotic cells, biological motors (e.g., kinesins) load/unload particular types of cargoes (e.g., vesicles) without external stimuli and transport them along cytoskeletal filaments (e.g., microtubules (MTs)) using the energy of adenosine triphosphate (ATP) hydrolysis [5]. Because of these biological capabilities of autonomous loading/unloading and transporting of specified cargoes, there is considerable interest in incorporating kinesins and MTs into artificially-created transporters and actuators in nano- or cell-scale systems and applications [10].

The authors of this paper have proposed a molecular propagation system that uses the reverse geometry of MT motility on kinesins and DNA hybridization/strand exchange [11]. The proposed propagation system uses DNA hybridization/strand exchange to achieve autonomous loading/unloading of specified cargoes (e.g., vesicles encapsulating information molecules) at a sender/receiver and MT motility to transport the loaded cargoes from a sender to a receiver (Fig. 3).

In order to use the DNA hybridization/strand exchange, each gliding MT, cargo, and unloading site is labeled with different single-stranded DNAs (ssDNAs). Note

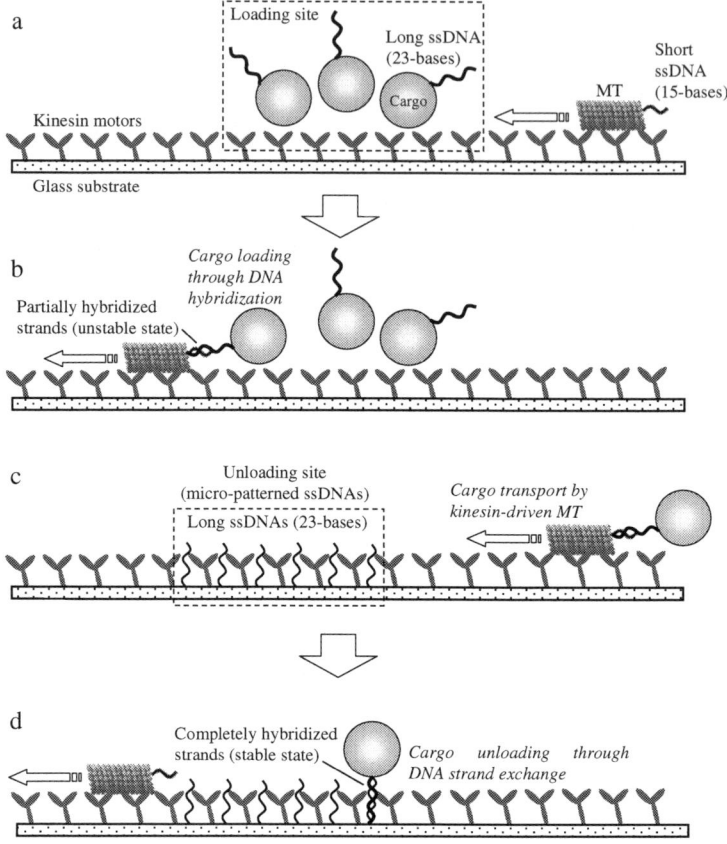

Fig. 3. A schematic diagram of a molecular propagation system using the reverse geometry of MT motility on kinesins and DNA hybridization/strand exchange

that the length of an ssDNA attached to an MT is designed to be shorter than that of the cargo, and the length of an ssDNA attached to a cargo is designed to be as long as that of the unloading site. Cargoes are pooled at a given loading site (a given sender) (Fig. 3a) and the ssDNA for the cargo is designed to be either complementary or non-complementary to that of the MT. When an MT labeled with an ssDNA passes through a given loading site, a cargo labeled with an ssDNA complementary to that of the MT is selectively loaded onto the gliding MT through DNA hybridization without external stimuli (Fig. 3b), while cargoes labeled with a non-complementary ssDNA remain at the loading site. The cargo loaded onto the MT (i.e., an MT-cargo complex) is transported by MT motility on kinesins toward a given unloading site (a given receiver) (Fig. 3c). To achieve autonomous unloading at a given unloading site, the ssDNA attached to each unloading site is designed to be either complementary or non-complementary to that attached to the cargo. When the MT-cargo complex passes through an unloading site, the cargo labeled with an ssDNA complementary to that attached to the unloading site is selectively unloaded from the gliding MT through DNA strand exchange without external stimuli (Fig. 3d).

In order to investigate the feasibility of the designed propagation system, the authors labeled MTs with ssDNAs using a chemical linkage that cross-links thiolated ssDNAs and amino groups of MTs, while maintaining smooth gliding of labeled MTs on kinesins [12]. Microscopic observations confirmed that 23-bases ssDNA labeled cargo-microbeads (used as model vesicles in which information molecules were encapsulated) were selectively loaded onto gliding MTs labeled with complementary 15-bases ssDNAs [12]. Microscopic observation also confirmed that loaded cargoes were selectively unloaded from the gliding MTs at a micro-patterned unloading site where complementary 23-bases ssDNAs were immobilized [13]. These results indicate that gliding MTs may load/unload cargo-vesicles at a sender/receiver through the DNA hybridization/strand exchange.

3.3 Receiver

A receiver selectively receives transported and decapsulated information molecules, and biochemically reacts to the received information molecules. Researchers at NAIST (Nara Institute of Science and Technology) worked with the authors of this paper and have proposed a receiver that uses a giant liposome embedded with gemini-peptide lipids [14]-[15]. A liposome is an artificially created vesicle that has the lipid bilayer membrane structure similar to vesicles and cells. The gemini-peptide lipids are composed of two amino acid residues, each having a hydrophobic double-tail and a functional spacer unit connecting to the polar heads of the lipid. The liposomes embedded with the same type of gemini-peptide lipids in their lipid bilayer membranes assemble in response to an external stimulus (e.g., light, ions, and temperature) [16]-[17]. This allows a selective reception of information molecules at a receiver (Fig. 4).

The gemini-peptide lipids are used as a molecular tag. A small liposome embedded with a molecular tag acts as a container of information molecules (a molecular container) and a giant liposome embedded with a molecular tag act as a receiver. A receiver is embedded with a specific molecular tag and a molecular container whose destination is the receiver is also embedded with the same type of molecular tag. When an external stimulus is applied to the receivers and the molecular containers, a receiver embedded with a molecular tag that is responsive to the applied external stimulus receives molecular containers embedded with the same type of molecular tag. This selective reception mechanism controlled by external stimuli may lead to creation of not only unicast-type but also multicast- and broadcast-type molecular communication.

In order to investigate the feasibility of the designed receiver, researchers at NAIST created molecular containers (liposomes with a diameter approximately 100 nm) and receivers (liposomes with a diameter larger than 2 μm) both containing zinc-ion responsive molecular tags in their lipid bilayer membranes. When zinc ions were added to the aqueous environment where both the molecular containers and receivers exist, the selective binding of the molecular containers to the receivers was observed [14]. The researchers at NAIST also created molecular containers and receivers both containing photo-responsive molecular tags in their lipid bilayer membranes to control reception of molecular containers to a receiver [15]. The photo-responsive molecular tag embedded in the receiver also acted as an artificial receptor, and interacted with an enzyme (signal amplifier) embedded in the receiver through metal ions when

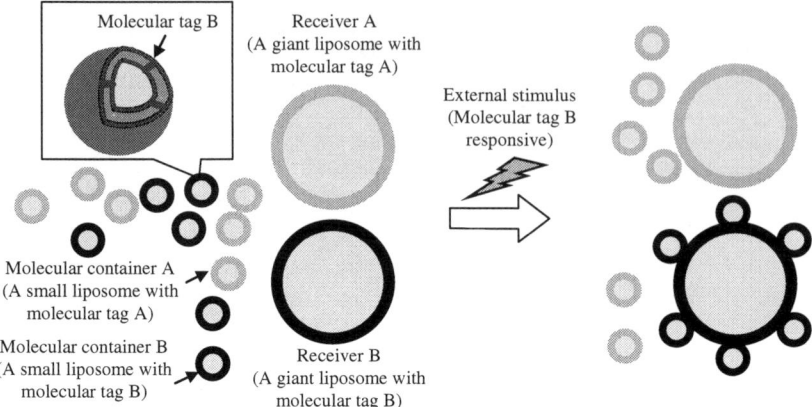

Fig. 4. A schematic diagram of receivers using giant liposomes with gemini-peptide lipids

a photonic signal was applied; achieving the signal amplification at the receiver by applying a photonic signal [15]. These results indicate that a receiver may selectively receive tagged molecular containers (encapsulating information molecules) and may biochemically react to the received information molecules by applying an external stimulus.

3.4 Integrated Molecular Communication System

The above sections have described system components in molecular communication (i.e., a molecular communication interface using a vesicle embedded with gap junction proteins, a propagation system using the MT motility on kinesins and DNA hybridization/strand exchange, and a receiver using a giant liposome embedded with gemini-peptide lipids). Described system components are compatible with each other and will be integrated into a single system (Fig. 5). Note that assembled liposomes

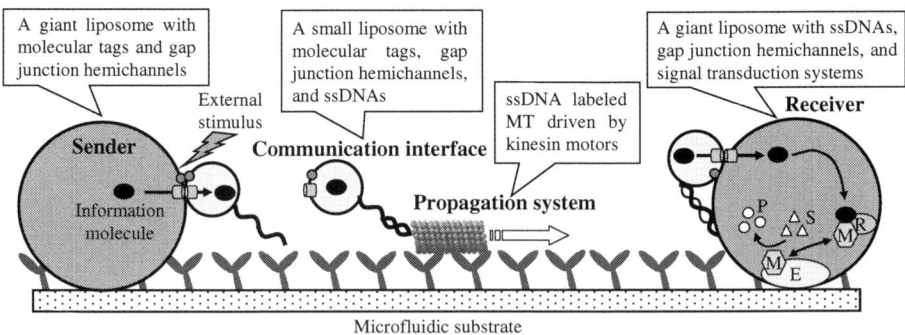

Fig. 5. A schematic diagram of an integrated molecular communication system. R, M, E, S, and P represent a receptor, a mediator, an enzyme, substrates, and products, respectively.

with molecular tags (gemini-peptide lipids) can be dissociated reversibly by applying a complementary external stimulus (e.g., applying UV-light for liposome assembly and applying visible light for liposome dissociation), and the selective reception mechanism may be applied to the selective transmission mechanism of molecular containers (small liposomes) at a sender.

4 Conclusions

This paper described basic concepts and key system components of molecular communication. This paper also discussed in detail system design of a communication interface that uses a vesicle embedded with gap junction proteins, a propagation system that uses MT motility on kinesins and DNA hybridization/strand exchange, and a receiver that uses a giant liposome with gemini-peptide lipids. The feasibility of the designed system components was confirmed through the biochemical experiments.

Molecular communication is an emerging interdisciplinary research area and is receiving increasing attention in the areas of biophysics, biochemistry, information science, and communication engineering [18]-[21]. The authors of this paper hope that a number of researchers participate in and contribute to the development of molecular communication.

Acknowledgments. The authors of this paper would like to acknowledge collaborators; Prof. Kazunari Akiyoshi and Associate Prof. Yoshihiro Sasaki (Tokyo Medical and Dental University), Prof. Kazuo Sutoh and Associate Prof. Shoji Takeuchi (The University of Tokyo), Prof. Jun-ichi Kikuchi (Nara Institute of Science and Technology). The authors also would like to thank Prof. Ikuo Morita (Tokyo Medical and Dental University), Dr. Shin-ichiro M. Nomura (Kyoto University), Prof. Akira Suyama and Prof. Yoko. Y. Toyoshima (The University of Tokyo) for their help with experiments described in this paper.

References

1. Hiyama, S., Moritani, Y., Suda, T., Egashira, R., Enomoto, A., Moore, M., Nakano, T.: Molecular Communication. In: Proc. NSTI Nanotechnology Conference and Trade Show, Anaheim, vol. 3, pp. 391–394 (May 2005)
2. Moritani, Y., Hiyama, S., Suda, T.: Molecular Communication - A Biochemically-Engineered Communication System. In: Proc. Frontiers in the Convergence of Bioscience and Information Technologies, Jeju Island, pp. 839–844 (October 2007)
3. Alberts, B., Bray, D., Johnson, A., Lewis, J., Raff, M., Roberts, K., Walter, P.: Essential Cell Biology – An Introduction to the Molecular Biology of the Cell. Garland Publishing (1998)
4. Luisi, P.L., Walde, P.: Giant Vesicles. John Wiley & Sons Inc., Chichester (2000)
5. Vale, R.D.: The Molecular Motor Toolbox for Intracellular Transport. Cell 112, 467–480 (2003)
6. Moritani, Y., Hiyama, S., Suda, T.: Molecular Communication among Nanomachines Using Vesicles. In: Proc. NSTI Nanotechnology Conference and Trade Show, Boston, vol. 2, pp. 705–708 (May 2006)

7. Moritani, Y., Nomura, S.-M., Hiyama, S., Akiyoshi, K., Suda, T.: A Molecular Communication Interface Using Liposomes with Gap Junction Proteins. In: Proc. Bio Inspired Models of Network, Information and Computing Systems, Cavalese (December 2006)
8. Kumar, N.M., Gilula, N.B.: The Gap Junction Communication Channel. Cell 84, 381–388 (1996)
9. Moritani, Y., Nomura, S.-M., Hiyama, S., Suda, T., Akiyoshi, K.: A Communication Interface Using Vesicles Embedded with Channel Forming Proteins in Molecular Communication. In: Proc. Bio Inspired Models of Network, Information and Computing Systems, Budapest (December 2007)
10. Van den Heuvel, M.G.L., Dekker, C.: Motor Proteins at Work for Nanotechnology. Science 317, 333–336 (2007)
11. Hiyama, S., Isogawa, Y., Suda, T., Moritani, Y., Sutoh, K.: A Design of an Autonomous Molecule Loading/Transporting/Unloading System Using DNA Hybridization and Biomolecular Linear Motors. In: Proc. European Nano Systems, Paris, pp. 75–80 (December 2005)
12. Hiyama, S., Inoue, T., Shima, T., Moritani, Y., Suda, T., Sutoh, K.: Autonomous Loading/Unloading and Transport of Specified Cargoes by Using DNA Hybridization and Biological Motor-Based Motility. Small 4, 410–415 (2008)
13. Hiyama, S., Takeuchi, S., Gojo, R., Shima, T., Sutoh, K.: Biomolecular Motor-Based Cargo Transporters with Loading/Unloading Mechanisms on a Micro-Patterned DNA Array. In: Proc. IEEE International Conference on Micro Electro Mechanical Systems, Tucson, pp. 144–147 (January 2008)
14. Sasaki, Y., Hashizume, M., Maruo, K., Yamasaki, N., Kikuchi, J., Moritani, Y., Hiyama, S., Suda, T.: Controlled Propagation in Molecular Communication Using Tagged Liposome Containers. In: Proc. Bio Inspired Models of Network, Information and Computing Systems, Cavalese (December 2006)
15. Mukai, M., Maruo, K., Kikuchi, J., Sasaki, Y., Hiyama, S., Moritani, Y., Suda, T.: Propagation and Amplification of Molecular Information Using a Photo-Responsive Molecular Switch. In: Proc. International Symposium on Macrocyclic & Supramolecular Chemistry, Las Vegas (July 2008)
16. Iwamoto, S., Otsuki, M., Sasaki, Y., Ikeda, A., Kikuchi, J.: Gemini peptide lipids with ditopic ion-recognition site. Preparation and functions as an inducer for assembling of liposomal membranes. Tetrahedron. 60, 9841–9847 (2004)
17. Otsuki, M., Sasaki, Y., Iwamoto, S., Kikuchi, J.: Liposomal sorting onto substrate through ion recognition by gemini peptide lipids. Chemistry Letters 35, 206–207 (2006)
18. Panel at IEEE INFOCOM 2005 (2005),
    http://www.ieee-infocom.org/2005/panels.htm
19. Symposium at EABS & BSJ 2006 (2006),
    http://www.ics-inc.co.jp//eabs2006/symposia.html#s3e2
20. Workshop at BIONETICS 2007 (2007),
    http://www.bionetics.org/2007/ccbs.shtml
21. NSF workshop,
    http://netresearch.ics.uci.edu/mc/nsfws08/index.html

Random Walks on Random Graphs

Colin Cooper[1] and Alan Frieze[2,*]

[1] Department of Computer Science, King's College, University of London,
London WC2R 2LS, UK
[2] Department of Mathematical Sciences, Carnegie Mellon University,
Pittsburgh PA15213, USA

1   Introduction

The aim of this article is to discuss some of the notions and applications of random walks on finite graphs, especially as they apply to random graphs. In this section we give some basic definitions, in Section 2 we review applications of random walks in computer science, and in Section 3 we focus on walks in random graphs.

Given a graph $G = (V, E)$, let $d_G(v)$ denote the degree of vertex v for all $v \in V$. The random walk $\mathcal{W}_v = (\mathcal{W}_v(t), t = 0, 1, \ldots)$ is defined as follows: $\mathcal{W}_v(0) = v$ and given $x = \mathcal{W}_v(t)$, $\mathcal{W}_v(t + 1)$ is a randomly chosen neighbour of x.

When one thinks of a random walk, one often thinks of Polya's Classical result for a walk on the d-dimensional lattice Z^d, $d \geq 1$. In this graph two vertices $x = (x_1, x_2, \ldots, x_d)$ and $y = (y_1, y_2, \ldots, y_d)$ are adjacent iff there is an index i such that (i) $x_j = y_j$ for $j \neq i$ and (ii) $|x_i - y_i| = 1$. Polya [33] showed that if $d \leq 2$ then a walk starting at the origin returns to the origin with probability 1 and that if $d \geq 3$ then it returns with probability $p(d) < 1$. See also Doyle and Snell [22].

A random walk on a graph G defines a Markov chain on the vertices V. If G is a finite, connected and non-bipartite graph, then this chain has a stationary distribution π given by $\pi_v = d_G(v)/(2|E|)$. Thus if $P_v^{(t)}(w) = \mathbf{Pr}(\mathcal{W}_v(t) = w)$, then $\lim_{t \to \infty} P_v^{(t)}(w) = \pi_w$, independent of the starting vertex v.

In this paper we only consider finite graphs, and we will focus on two aspects of a random walk: The *Mixing Time* and the *Cover Time*.

1.1   Mixing Time

For $\epsilon > 0$ let

$$T_G(\epsilon) = \max_v \min \left\{ t : ||P_v^{(t)} - \pi||_{TV} \leq \epsilon \right\},$$

where

$$||P_v^{(t)} - \pi||_{TV} = \frac{1}{2} \sum_w |P_v^{(t)}(w) - \pi_w|$$

is the Total Variation distance between $P_v^{(t)}$ and π.

* Supported in part by NSF grant CCF0502793.

M. Cheng (Ed.): NanoNet 2008, LNICST 3, pp. 95–106, 2009.

We say that a random walk on G is *rapidly mixing* if $T_G(1/4)$ is $poly(\ln|V|)$, where $\ln = \log_e$ is the natural logarithm. The choice of $1/4$ is somewhat arbitrary, any constant strictly less than $1/2$ will suffice. Rapidly mixing Markov chains are extremely useful and we will have more to say on them in Sections 2.1.1 – 2.1.3.

1.2   Cover Time

For $v \in V$ let C_v be the expected time taken for a simple random walk W on G starting at v, to visit every vertex of G. The *vertex cover time* C_G of G is defined as $C_G = \max_{v \in V} C_v$. The (vertex) cover time of connected graphs has been extensively studied. It is a classic result of Aleliunas, Karp, Lipton, Lovász and Rackoff [4] that $C_G \leq 2m(n-1)$. It was shown by Feige [24], [25], that for any connected graph G, the cover time satisfies $(1 - o(1))n \ln n \leq C_G \leq (1 + o(1))\frac{4}{27}n^3$. As an example of a graph achieving the lower bound, the complete graph K_n has cover time determined by the Coupon Collector problem. The *lollipop* graph consisting of a path of length $n/3$ joined to a clique of size $2n/3$ gives the asymptotic upper bound for the cover time. We will have more to say on the cover time in Sections 2.2 and 3.

2   Applications in Computer Science

2.1   Rapid Mixing

2.1.1   Sampling and Counting

Let Δ denote the maximum degree of a graph $G = (V, E)$. Suppose now that we wish to find a *uniform random* proper colouring of G using $k \geq \Delta + 1$ colours. By proper we mean that adjacent vertices get different colours. We can easily generate one such colouring via a greedy algorithm, but it will certainly not be a random colouring. The distribution of random colourings of G is complex. The only known approach to this sampling problem is via random walk. Let Ω denote the set of all proper k colourings of G. This will most likely be of exponential size in $n = |V|$. Now consider an auxilliary *multi-graph* $\Gamma = (\Omega, F)$. Two distinct colourings $\sigma_1, \sigma_2 : V \to [k]$ are adjacent if they only differ at one vertex. In addition we add self-loops to make every vertex have the same degree. A random walk on Γ is equivalent to the following Markov chain $X_0, X_1, \ldots$, on Ω: Given X_t we generate X_{t+1} as follows:

1. Choose z uniformly at random from V, and c uniformly at random from $\{1, \ldots, k\}$.
2. For all $w \neq z$, set $X_{t+1}(z) = X_t(z)$.
3. If no neighbors of z have color c (i.e., $c \notin X_t(N(z))$), then set $X_{t+1}(z) = c$, otherwise set $X_{t+1}(z) = X_t(z)$.

The version above is called *Metropolis Dynamics*.

Jerrum [29] showed that if $k \geq 2\Delta$ then this chain/walk is rapidly mixing i.e. it gets close to the uniform distribution in time $poly(n)$. From here it is straightforward to devise an algorithm that gives a good approximation to $|\Omega|$ in polynomial time, see [29].

Counting colourings is only a single example of a flourishing research area involving the use of Markov chains to sample from complex distributions and to estimate the size of large combinatorially defined sets. This area of research constitutes an important meeting place for researchers in Statistical Physics and Theoretical Computer Science. For further reading, see [26] or [30].

2.1.2 Expanders

In this section we will for convenience asssume that G is a d-regular graph i.e. every vertex has the same degree d. The adjacency matrix $\mathbf{A}$, where $\mathbf{A}(v, w) = 1$ iff v, w are adjacent. $\mathbf{A}$ has largest eigenvalue d and suppose now that λ is its second largest eigenvalue. In this case it is known that

$$T_G(\epsilon) \leq \lceil \frac{\ln \epsilon n}{\ln \lambda} \rceil. \tag{1}$$

We will say that a graph G is an α-expander if for all $S \subseteq V$ with $|S| \leq n/2$, we have $e(S : \bar{S}) \geq \alpha |S|$ where $e(S : \bar{S})$ is the number of edges from S to $\bar{S} = V \setminus S$. In which case one can show that $\lambda \leq 1 - \frac{\alpha^2}{2d^2}$. So if $d = O(1)$ and $\alpha = \Omega(1)$ then a random walk on G mixes in $O(\ln n)$ time. (See Alon [5] and Jerrum and Sinclair [31]. This property of expanders can be used to reduce the number of random bits needed by a randomized algorithm.

We explain the use of the following theorem of Ajtai, Komlós and Szemerédi [1]:

Theorem 1. *Let $G = (V, E)$ be a d-regular graph on n. Let C be a set of cn vertices of G. Then for every ℓ, the number of walks of length ℓ in G that avoid C does not exceed $(1 - c)n((1 - c)d + c\lambda)^\ell$.*

This means that a random walk of length ℓ, with a randomly chosen start vertex, has probability at most $(1 - c)(1 - c(1 - \lambda/d))^\ell < e^{-c(1-\lambda/d)\ell}$ of avoiding C completely. In this context, consider the Miller-Rabin algorithm for testing whether an integer n is prime. Without going into details, it is known that if n is composite then at least $1/2$ of the integers between 1 and n can be used to verify this. So if n is composite and we choose ℓ random integers between 1 and n then the probability we fail to show it is composite is at most $2^{-\ell}$. If we choose our random integers in the normal way, then this requires $\ell \ln_2 n$ random bits. On the other hand, suppose that we have a d-regular graph on $[n]$ with $\lambda \leq \epsilon d$ say, and we do a random walk of length ℓ from a randomly chosen start, then this requires $\ln_2 n + \ell \ln_2 d$ random bits. A significant saving if $d = O(1)$. Applying Theorem 1 we see that the probability we fail to show it is composite is at most $\left(\frac{1+\epsilon}{2}\right)^\ell$.

There are many uses of expanders. See Hoory, Linial and Wigderson [28] for a survey.

2.1.3 Edge Disjoint Paths

In this section we discuss the use of random walks to find edge disjoint paths in an expander graph G. We are given a graph $G = (V, E)$ with n vertices, and a set of κ pairs of vertices in V, we are interested in finding for each pair (a_i, b_i), a path connecting a_i to b_i, such that the set of κ paths so found is edge-disjoint. In general this is an NP-hard problem, but some strong results have been proven in the context of d-regular expanders. In a series of papers, culminating in Frieze [27] it was shown that if G is a sufficiently strong expander and if $\epsilon > 0$ is a sufficently small constant then this problem can be solved if $\kappa \leq \epsilon n / \ln n$. First the edge set of G is partitioned to create several edge disjoint expander graphs $G_1, G_2, \ldots G_{10}$. Various phases of the algorithm take place on various G_i. a network flow algorithm is used to connect the a_i to randomly chosen κ-set of vertices by edge disjoint paths. The other endpoint of the path with one endpoint a_i is labelled a_i'. In a similar manner b_i is connected to a randomly chosen b_i'. Finally, a path $a_i', \ldots, x_i, \ldots, b_i'$ is constructed. Here x_i is chosen according to the steady state of a random walk on one of the G_i's and then both a_i' and b_i' are connected to x_i via a random walk.

2.1.4 Randomized Dual Simplex Algorithm

In this section we discuss the use of random walks to solve the linear program

LP(b) : minimise cy subject to $\mathbf{A}y = b$, $y \geq 0$. where A is a totally unimodular matrix i.e. all sub-determinants of $\mathbf{A}$ are $0, \pm 1$.

The algorithm described in Dyer and Frieze [23] is somewhat complicated, but it can be viewed as a dual simplex algorithm for the above problem, in which the choice of next pivot is found via a random walk through a geomtrically defined graph. The reader is refered to the paper for details.

2.2   Cover-Time

2.2.1   Log-Space Algorithm for $s - t$ Connecitivity

One of the earliest computer science applications of random walk is in [4]. The problem under discussion was whether it is possible to check whether two vertices s and t are in the same component of a bounded degree graph G. The time constraint is polynomial, but the space constraint is logarithmic i.e. only $O(\ln n)$ working storage is allowed. This rules out algorithms like breadth-first and depth-first search. The solution was to consider a random walk from s and run it for $O(mn)$ time. This only requires order $\ln n$ storage and one of the main results of the paper was that the cover time C_G of a connected graph satisfies $C_G \leq 2m(n-1)$. Suppose then that we repeat the following ℓ times and still do not reach t. Take a random walk of length $4mn$ from s. If s, t are in the same component then this walk will go through t with probability at least $1 - \frac{2m(n-1)}{4mn} \geq \frac{1}{2}$. So if s, t are indeed in the same component then this algorithm succeeds with probability $\geq 1 - 2^{-\ell}$. It is only recently that Reingold [34] has found a deterministic algorithm that uses LOGSPACE.

2.2.2   Universal Traversal Sequences

Suppose that G is a d-regular graph G and that for each vertex $v \in V$ we order the neighbours of v as $x(v, i)$, $i = 1, 2, \ldots, d$. We call this an ordered d-regular graph. Given a start vertex v and a sequence $\sigma = (i_1, i_2, \ldots, i_\ell) \in [d]^\ell$, $\ell = |\sigma|$ we can define a walk $P(v, \sigma) = (v = y_0, y_1, \ldots, y_\ell)$ by $y_{j+1} = x(y_j, i_j)$ i.e. y_{j+1}

is the i_jth neighbour of y_j. $P(v, \sigma)$ traverses G if it visits each vertex of G. If we choose σ at random from $[d]^{\ell}$ and v arbitrarily then the walk $P(v, \sigma)$ is a random walk. Let Z_v denote the time taken by the random walk $\mathcal{W}_v$ to visit all vertices of G. We see then that if $|\sigma| = 4m(n-1)$ then

$$\mathbf{Pr}(\sigma \text{ does not traverse } G) = \mathbf{Pr}(Z_v \geq 4m(n-1)) \leq \mathbf{Pr}(Z_v \geq 2\mathbf{E}(Z_v)) \leq \tfrac{1}{2}.$$

Similarly, if $|\sigma| = 4km(n-1)$ then

$$\mathbf{Pr}(\sigma \text{ does not traverse } G) \leq \tfrac{1}{2^k}.$$

Now there are at most n^{dn+1} ways of choosing an ordered d-regular graph and a start vertex. So, with $|\sigma| = 4km(n-1)$,

$$\mathbf{Pr}(\exists \text{ordered } d\text{-regular graph } G \text{ and start vertex } v \text{ such that } \sigma \text{ does not traverse } G) \leq \frac{n^{dn+1}}{2^k}.$$

$$(2)$$

If $k > (dn+1)\ln_2 n$ then the RHS of (2) is less than one. Thus there exists a sequence σ of length $O(dmn \ln n)$ such that for every ordered G and every start vertex v, $P(v, \sigma)$ traverses G. Put another way, using this *universal traversal sequence* we can be sure of ariving at any other vertex, if we follow σ. This being regardless of start vertex v and graph G. So, short (polynomial length) universal traversal sequences exist, but they are very hard to construct explicitly.

2.2.3   Random Spanning Trees

Aldous [2] and Broder [10] independently proved the following beautiful result concerning spanning trees of a fixed graph G. Initialise $T = \emptyset$ and start a random walk $\mathcal{W}$ at an arbitrary vertex and when the walk first visits a vertex w add the edge (v, w) to T. Here (v, w) is the edge just traversed by $\mathcal{W}$. The algorithm stops after $\mathcal{W}$ has visited all vertices. The algorithm generates a spanning tree T of G. The aforementioned papers prove that T is equally likely to be any spanning tree of G. A rather remarkable result.

3   Random Graphs

Various topics arise in the context of random walks on random graphs. Among them are the following: Mixing time of the random walk, cover time of a random graph, properties of multiple particle walks, random walks on graph processes, constructing random networks using random walks.

3.1   Mixing Time

There is not much to say here except that random graphs tend to be excellent expanders. In some sense they provide the simplest method of generating an expander graph. For example a random walk on an r-regular graph, $r \geq 3$, has mixing time $O(\log n)$ **whp**. In contrast it has proven very difficult to produce *explicit* expanders. It has been done, but the methods can be deep and complicated.

3.2   Cover Time of Random Graphs

In this section we study the cover time of various classes of random graphs with fixed vertex set $V = [n] = \{1, 2, ..., n\}$. The spaces of labeled random graphs we consider here are: Erdos-Renyi graphs $G_{n,p}$, random digraphs $D_{n,p}$, random r-regular graphs $\mathcal{G}_r$, preferential attachment graphs $G_m(n)$ and random geometric graphs $G = G(d, r, n)$. A fuller definition of these graph spaces is given below.

It is probably a good time to mention a graph of particular interest in the context of this meeting i.e. *Carbon Nano-Tube* networks, see for example Bush and Li [12]. These graphs are formed from the intersection points of "randomly" placed line segments and one is interested in their *effective resistance*. This parameter is related to commute times, which are related to cover time. We do not have any results yet on a model of such graphs, but it forms a promising line of research.

A few words on notation. Results on random graphs are always asymptotic in n, the size of the vertex set. The notation $A_n \sim B_n$ means that $\lim_{n\to\infty} A_n/B_n = 1$, and **whp** (with high probability) means with probability tending to 1 as $n \to \infty$.

Erdos-Renyi graphs $G_{n,p}$ are defined as follows. The edge $\{i, j\}$ between any pair of vertices i and j occurs with probability p, independently of all other edges. Let C_G denote the vertex cover time. It was shown by Jonasson [32] that **whp**

- $C_G = (1 + o(1))n \ln n$ if $\frac{np}{\ln n} \to \infty$.
- If $c > 1$ is constant and $np = c \ln n$ then $C_G > (1+\alpha)n \ln n$ for some constant $\alpha = \alpha(c)$.

Thus Jonasson has shown that when the expected average degree $(n-1)p$ grows faster than $\ln n$, a random graph has the same cover time **whp** as the complete graph K_n, whose cover time is determined by the Coupon Collector problem. Whereas, when $np = \Omega(\ln n)$ this is not the case. This result was refined for sparse graphs as follows:

- If $p = d \ln n/n$ and $d > 1$ then **whp** $C_{G_{n,p}} \sim d \ln \left(\frac{d}{d-1}\right) n \ln n$, [16].
- Let $d > 1$ and let x denote the solution in $(0, 1)$ of $x = 1 - e^{-dx}$. Let X_g be the giant component of $G_{n,p}$, $p = d/n$. Then **whp** $C_{X_g} \sim \frac{dx(2-x)}{4(dx-\ln d)} n(\ln n)^2$, [17]

Considering random r-regular graphs (i.e. the set of all simple r-regular graphs with the uniform measure), we have the following result [14]:

If $G_{n,r}$ denotes a random r-regular graph on vertex set $[n]$ with $r \geq 3$ then **whp** $C_{G_{n,r}} \sim \frac{r-1}{r-2} n \ln n$.

The proof of this result uses a lemma, which we call the first visit time lemma, which under not very restrictive conditions (see e.g. [17]) states that the probability $f(v; T, ..., t)$ that vertex v is not visited by the walk during steps $T, ..., t$ is given by

$$f(v; T, ..., t) = (1 + o(1))(1 - p_v)^t,$$

where T is a mixing time of the walk. Here $p_v \sim \frac{\pi_v}{R_v}$, where π_v is the stationary distribution of vertex v, and R_v is the expected number of returns to v during the mixing time T, of a random walk starting at v. Thus R_v is dependent only on the local geometry of the graph around v. This result is true also for weighted random walks, and general ergodic Markov processes.

The preferential attachment graph $G_m(n)$ is a random graph formed by adding a new vertex at each time step, with m edges which point to vertices selected at random with probability proportional to their degree. Thus at time n there are n vertices and mn edges. This process yields a graph which has been proposed as a simple model of the world wide web [8]. In [15] it is shown that if $m \geq 2$ then **whp** $C_{G_m(n)} \sim \frac{2m}{m-1} n \ln n$.

The random digraphs $D_{n,p}$ are generated in the same manner as $G_{n,p}$ except that now, each directed edge (i,j) occurs independently with probability p. The first visit time lemma applies to these graphs provided they are strongly connected (etc) and we find that: If $p = d \ln n / n$ and $d > 1$ then **whp** $C_{D_{n,p}} \sim d \ln \left(\frac{d}{d-1} \right) n \ln n$.

The main problem for walks on directed graphs is to determine the stationary distribution π_v.

Finally we consider geometric random graphs. Let I denote the unit interval $[0,1]$ and let $I(d) = [0,1]^d$ denote the unit torus in d dimensions. We define a random geometric graph $G = G(d,r,n)$ as follows: Sample n points V independently and uniformly at random from $I(d)$ wrapped toroidally. For each point x draw a ball (disk) $D(x,r)$ of radius r about x. The vertex set $V(G) = V$ and the edge set $E(G) = \{\{v,w\} : w \neq v, w \in D(v,r)\}$. The graph serves as a model for ad-hoc networks, where transmitters have limited range.

Avin and Ercal [6] considered the case $d = 2$. They proved that if $G = G(2,r,n)$ and $r^2 > (8 \ln n / n)$ then **whp**$C_G = \Theta(n \ln n)$. For $d \geq 3$ dimensions we can give precise results. Let $G(d,r,n)$, $d \geq 3$ be a random geometric graph. Let $r = (c \ln n)/(\Upsilon_d n)^{1/d}$ and where $c > 1$ is a constant. Then **whp**

$$C_G \sim c \ln \left(\frac{c}{c-1} \right) n \ln n. \tag{3}$$

Here $\Upsilon_d = (\pi^{d/2})/\Gamma(d/2 + 1)$ is the volume of the unit ball $D(0,1)$ in d dimensions.

3.3   Multiple Particle Walks

Suppose there are $k \geq 1$ particles, each making a simple random walk on a graph G. Essentially there are two possibilities. Either the particles are *Oblivious* or *Interactive*. Oblivious particles act independently of each other, and do not interact on meeting. They may however interact with vertices, possibly in a way determined by previous visits of other particles. Interactive particles, can interact directly in some way on meeting. We assume that interaction only occurs when meeting at a vertex, and that the random walks made by the particles are otherwise independent. Various models and questions arise, e.g.

- **Multiple walks.** For k particles walking independently, we establish the cover time $C_G(k)$ of G.
- **Talkative particles.** For k particles walking independently, which communicate on meeting at a vertex, we study the expected time to broadcast a message.

- **Predator-Prey.** For k predator and ℓ prey particles walking independently, we study the expected time to extinction of the prey particles, when predators eat prey particles on meeting at a vertex.
- **Coalescing particles.** For k particles walking independently, which coalesce on meeting at a vertex, we study the expected time to coalesce to a single particle.
- **Annihilating particles.** For $k = 2\ell$ particles walking independently, which destroy each other (pairwise) on meeting at a vertex, we study the expected time to extinction.

The motivation for these models comes from many sources. Using random walks to test graph connectivity is an established algorithm, and it is appealing to speed up this by parallel searching [11], [7]. Similarly, properties of communication, such as broadcasting and gossiping, between particles moving in a network, is a natural question. In this context, the predator-prey model represents interaction between server and client particles, where each client needs to attach to a server.

Coalescing and annihilating particle systems are part of the classical theory of interacting particles (see e.g. [3]). A system of coalescing particles where initially one particle is located at each vertex, corresponds to another classical problem, the voter model, which is defined as follows: Initially each vertex has a distinct opinion, and at each step each vertex changes its opinion to that of a random neighbour. It is known that the expected time for a unique opinion to emerge, is the same as the expected time for all the particles to coalesce. By establishing the expected coalescence time, we obtain the expected time for voting to be completed.

The cover time of a random walk on a random r-regular graph was studied in [14], where it was shown with high probability (**whp**), that for $r \geq 3$ the cover time is asymptotic to $\theta_r n \ln n$, where $\theta_r = (r-1)/(r-2)$.

In [20] we prove the following (**whp**) results, arising from the study of multiple random walks on a random regular graph G. For k independent walks on G, the cover time $C_G(k)$ is asymptotic to C_G/k, where C_G is the cover time of a single walk. For most starting positions, the expected number of steps before any of the walks meet is $\theta_r n / \binom{k}{2}$. If the walks can communicate when meeting at a vertex, we show that, for most starting positions, the expected time for k walks to broadcast a single piece of information to each other is asymptotic to $\frac{2 \ln k}{k} \theta_r n$, as $k, n \to \infty$.

We also establish properties of walks where there are two types of particles, predator and prey, or where particles interact when they meet at a vertex by coalescing, or by annihilating each other. For example, the expected coalescence time of k particles tends to $2\theta_r n$ as $k \to \infty$; the expected extinction time of k explosive particles (k even) tends to $(2 \ln 2)\theta_r n$ as $k \to \infty$. Suppose k predator and ℓ prey particles make random walks, starting in general position (not too near each other). Let $D_{k,\ell}$ be the extinction time of the prey. Then $\mathbf{E}(D_{k,\ell}) \sim \frac{\theta_r H_\ell}{k} n$.

The case of n coalescing particles, where one particle is initially located at each vertex, corresponds to a voter model defined as follows: Initially each vertex has a distinct opinion, and at each step each vertex changes its opinion to that of a random neighbour. The expected time for a unique opinion to emerge is the expected time for all the particles to coalesce, which is asymptotic to $2\theta_r n$.

Combining results from the predator-prey and multiple random walk models allows us to compare expected detection time in the following scenarios: both the predator and the prey move randomly, the prey moves randomly and the predators stay fixed, the predators move randomly and the prey stays fixed. In all cases, with k predators and ℓ prey the expected detection time is $\theta_r H_\ell n/k$, where H_ℓ is the ℓ-th harmonic number. A application of this is with the predators as government agents and the prey as criminals.

3.4 Random Walks on Random Graph Processes

If we consider a random graph process $(G(t),\ t = 0, 1, ...)$ in which the graph evolves at each step by the addition of vertices and/or edges then the random walk is searching a growing graph, so we cannot hope to visit all vertices of the graph.

For example, consider a simple model of search, on e.g. the WWW, in which a particle (which we call a spider) makes a random walk on the nodes of an undirected graph process. It is presumed that the spider examines the data content of the nodes for some specific topic. As the spider is walking the graph is growing, and the spider makes a random transition to whatever neighbours are available at the time. For simplicity, we assume that the growth rate of the process and the transition rate of the random walk are similar, so that the spider has at least a chance of crawling a constant proportion of the process. Although the edges of the WWW graph are directed, the idea of evaluating models of search on an undirected process has many attractions, not least its simplicity.

We study the success of the spider's search on comparable graph processes of two distinct types: a random graph process and a web graph process [13]. In the simple process we consider, each new vertex directs m edges towards existing vertices, either choosing vertices randomly (giving a random graph process) or copying according to vertex degree (giving a web graph process). Once a vertex has been added the direction of the edges is ignored.

We consider the following models for the graph process $G(t)$. Let $m \geq 1$ be a fixed integer. Let $[t] = \{1, ..., t\}$ and let $G(1) \subset G(2) \subset \cdots \subset G(t)$. Initially $G(1)$ consists of a single vertex 1 plus m loops. For $t \geq 2$, $G(t+1)$ is obtained from $G(t)$ by adding the vertex t and m randomly chosen edges $\{t + 1, v_i\}, i = 1, 2, \ldots, m$, where

Model 1: Vertices $v_1, v_2, \ldots, v_m$ are chosen independently and uniformly with replacement from $[t]$.

Model 2: Vertices $v_1, v_2, \ldots, v_m$ are chosen proportional to their degree after step t. Thus if $d(v, \tau)$ denotes the degree of vertex v in $G(\tau)$ then for $v \in [t]$ and $i = 1, 2, \ldots, m$,

$$\mathbf{Pr}(v_i = v) = \frac{d(v, t)}{2mt}.$$

While vertex t is being added, the spider $\mathcal{S}$ is sitting at some vertex X_{t-1} of $G(t-1)$. After the addition of vertex t, and before the beginning of step $t+1$, the spider now makes a random walk of length ℓ, where ℓ is a fixed positive integer independent of t.

Let $\eta_{\ell,m}(t)$ be the expected proportion of vertices which have not been visited by the spider at step t, when t is large. If we allow $m \to \infty$ we can get precise asymptotic values. Let $\eta_\ell = \lim_{m \to \infty} \eta_{\ell,m}$, then

(a) For Model 1,

$$\eta_\ell = \sqrt{\frac{2}{\ell}} e^{(\ell+2)^2/(4\ell)} \int_{(\ell+2)/\sqrt{2\ell}}^{\infty} e^{-y^2/2}\, dy, \quad \eta_1 = 0.57 \cdots, \text{ and } \eta_\ell \sim 2/\ell \text{ as } \ell \to \infty.$$

(b) For Model 2

$$\eta_\ell = e^\ell 2\ell^2 \int_\ell^{\infty} y^{-3} e^{-y}\, dy, \quad \eta_1 = 0.59 \cdots, \text{ and } \eta_\ell \sim 2/\ell \text{ as } \ell \to \infty.$$

So for large m, t and $\ell = 1$ it is slightly harder for the spider to crawl on a webgraph whose edges are generated by a copying process (Model 2) than on a uniform choice random graph (Model 1).

3.5   Constructing Random Networks Using Random Walks

Bourassa and Holt [9] propose a decentralised protocol for P2P networks based on random walks. If a vertex in the network needs an address of a random vertex, then it initiates a random walk and gets the address of the vertex reached at some specified step of the walk. The protocol constructs a 4-regular random graph. Their protocol, however, cannot reconnect the network if it becomes disconnected.

In [21] we describe a randomized algorithm for assigning neighbours to vertices joining a P2P network. The aim of the algorithm is to maintain connectivity, low diameter and constant vertex degree. On joining each vertex donates a constant number c of tokens to the network. These tokens contain the address of the donor vertex. Tokens make independent random walks in the network. A token can be used by any vertex it is visiting, to establish a connection to the donor vertex. This allows vertices which initially join in an arbitrary manner (e.g to a friend/super-node) to be re-allocated to a random set of neighbours although the overall vertex membership of the network is unknown. The new vertex joins arbitrarily, collects m tokens, attaches to the vertices whose addresses they contain and detaches from its original contacts.

If t is the size of the network, then the diameter of the network is $O(\ln t)$ for all t, with high probability. The network is extremely robust under adversarial deletion of vertices and edges and actively reconnects itself when broken. As an example of the robustness of this model, suppose an adversary deletes edges from the network leaving components of size at least $t^{1/2+\delta}, \delta > 0$ small. With high probability the network rapidly reconnects itself by replacing lost edges using tokens from the token pool.

References

1. Ajtai, M., Komlós, J., Szemerédi: Deterministic simulation in LOGSPACE. In: Proceedings of the 19th Annual ACM Symposium on Theory of Computing, pp. 132–140 (1987)
2. Aldous, D.: The Random Walk Construction of Uniform Spanning Trees and Uniform Labelled Trees. SIAM Journal on Discrete Mathematics 3, 450–465 (1990)
3. Aldous, D., Fill, J.: Reversible Markov Chains and Random Walks on Graphs, http://stat-www.berkeley.edu/pub/users/aldous/RWG/book.html
4. Aleliunas, R., Karp, R.M., Lipton, R.J., Lovász, L., Rackoff, C.: Random Walks, Universal Traversal Sequences, and the Complexity of Maze Problems. In: Proceedings of the 20th Annual IEEE Symposium on Foundations of Computer Science, pp. 218–223 (1979)
5. Alon, N.: Eigenvalues and expanders. Combinatorica 6, 83–96 (1986)
6. Avin, C., Ercal, G.: On the cover time and mixing time of random geometric graphs. Theoretical Computer Science 380, 2–22 (2007)
7. Alon, N., Avin, C., Kouchý, M., Kozma, G., Lotker, Z., Tuttle, M.: Many random walks are faster then one (to appear)
8. Barabási, A., Albert, R.: Emergence of scaling in random networks. Science 286, 509–512 (1999)
9. Bourassa, V., Holt, F.: SWAN: Small-world wide area networks. In: Proceedings of International Conference on Advances in Infrastructure (SSGRR 2003w), L'Aquila, Italy, paper # 64 (2003), http://www.panthesis.com/content/swan_white_paper.pdf
10. Broder, A.Z.: Generating Random Spanning Trees. In: Proceedings of the 30th Annual IEEE Symposium on Foundations of Computer Science, pp. 442–447 (1989)
11. Broder, A., Karlin, A., Raghavan, A., Upfal, E.: Trading space for time in undirected s-t connectivity. In: Proc. ACM Symp. Theory of Computing, pp. 543–549 (1989)
12. Bush, S.F., Li, Y.: Network Characteristics of Carbon nanotubes: A Graph Eigenspectrum Approach and Tool Using Mathematica, GE Technical Information Series Report: 2006GRC023
13. Cooper, C., Frieze, A.: Crawling on simple models of web-graphs. Internet Mathematics 1, 57–90 (2003)
14. Cooper, C., Frieze, A.M.: The cover time of random regular graphs. SIAM Journal on Discrete Mathematics 18, 728–740 (2005)
15. Cooper, C., Frieze, A.M.: The cover time of the preferential attachment graph. Journal of Combinatorial Theory Series B 97(2), 269–290 (2007)
16. Cooper, C., Frieze, A.M.: The cover time of sparse random graphs. Random Structures and Algorithms 30, 1–16 (2007)
17. Cooper, C., Frieze, A.M.: The cover time of the giant component of a random graph. Random Structures and Algoritms 32, 401–439 (2008)
18. Cooper, C., Frieze, A.M.: The cover time of random digraphs. In: Charikar, M., Jansen, K., Reingold, O., Rolim, J.D.P. (eds.) RANDOM 2007 and APPROX 2007. LNCS, vol. 4627, pp. 422–435. Springer, Heidelberg (2007)
19. Cooper, C., Frieze, A.M.: The cover time of random geometric graphs (to appear)
20. Cooper, C., Frieze, A., Radzik, T.: Multiple random walks in random regular graphs (to appear)
21. Cooper, C., Klasing, R., Radzik, T.: A randomized algorithm for the joining protocol in dynamic distributed networks. J. Theoretical Computer Science (to appear); also published as INRIA report RR-5376 and CNRS report I3S/RR-2004-39-FR

22. Doyle, P.G., Snell, J.: Random Walks and Electric Networks, Mathematical Association of America (1984), http://xxx.lanl.gov/abs/math.PR/0001057
23. Dyer, M.E., Frieze, A.M.: Random walks, totally unimodular matrices and a randomised dual simplex algorithm. Mathematical Programming 64, 1–16 (1994)
24. Feige, U.: A tight upper bound for the cover time of random walks on graphs. Random Structures and Algorithms 6, 51–54 (1995)
25. Feige, U.: A tight lower bound for the cover time of random walks on graphs. Random Structures and Algorithms 6, 433–438 (1995)
26. http://www.math.cmu.edu/~af1p/Mixingbook.pdf
27. Frieze, A.M.: Edge disjoint paths in expander graphs. SIAM Journal on Computing 30, 1790–1801 (2001)
28. Hoory, S., Linial, N., Wigderson, A.: Expander graphs and their applications. Bulletin of the American Mathematical Society 43, 439–561 (2006)
29. Jerrum, M.R.: A very simple algorithm for estimating the number of k-colourings of a low-degree graph. Random Structures and Algorithms 7(2), 157–165 (1995)
30. Jerrum, M.R.: Counting, sampling and integrating: algorithms and complexity. Lectures in Mathematics–ETH Zürich, Birkhäuser (2003)
31. Jerrum, M., Sinclair, A.: The Markov chain Monte Carlo method: an approach to approximate counting and integration. In: Hochbaum, D. (ed.) Approximation Algorithms for NP-hard Problems, pp. 482–520. PWS (1996)
32. Jonasson, J.: On the cover time of random walks on random graphs. Combinatorics, Probability and Computing 7, 265–279 (1998)
33. Polya, G.: Über eine Aufgabe der Wahrscheinlichkeitstheorie betreffend die Irrfahrt im Strassennetz. Mathematische Annalen 84, 149–160 (1921)
34. Reingold, O.: Undirected ST-Connectivity in Log-Space. In: Proceedings of the 37th Annual ACM Symposium on Theory of Computing, pp. 376–385 (2005)

Optical Networking in a Swarm of Microrobots

Paolo Corradi[1], Thomas Schmickl[2], Oliver Scholz[3,4], Arianna Menciassi[1],
and Paolo Dario[1]

[1] Scuola Superiore Sant'Anna, Pisa, Italy
paolo.corradi@sssup.it
http://www.sssup.it/
[2] Department for Zoology, Karl-Franzens-University Graz, Graz, Austria
[3] Fraunhofer Institute for Biomedical Engineering, Sankt Ingbert, Germany
[4] University of Saarland, Saarbrcken, Germany

Abstract. Swarm Microrobotics aims to apply Swarm Intelligence algorithms and strategies to a large number of fabricated miniaturized autonomous or semi-autonomous agents, allowing collective, decentralized and self-organizing behaviors of the robots. The ability to establish basic information networking is fundamental in such swarm systems, where inter-robot communication is the base of emergent behaviors. Optical communication represents so far probably the only feasible and suitable solution for the constraints and requirements imposed by the development of a microrobotic swarm. This paper introduces a miniaturized optical communication module for millimeter-sized autonomous robots and presents a computer-simulated demonstration of its basic working principle to exploit bio-inspired swarm strategies.

Keywords: micro-optics, optical communication, microrobotics, swarm intelligence.

1 Introduction

Microrobotics is a field of the scientific and technological research that aims for the development of miniaturized autonomous or semi-autonomous systems. Microrobotics has a particular relevance in the development of a relatively new scientific discipline named *Swarm Robotics*. This aims to apply *Swarm Intelligence* strategies [1] to a large number of robotic agents, allowing collective, decentralized and self-organizing behaviours of the robots, possibly leading to a global, often bio-inspired, intelligent behavior of the swarm on the base of few simple basic rules, while considering issues of robustness and scalability. It is indeed in the perspective of miniaturization that Swarm Robotics becomes meaningful, leading to the concept of *Swarm Microrobotics*. Actually, microrobots have by construction very limited capabilities, thus they need to operate in very large groups, or swarms, in order to have any appreciable effect on the "macroworld". In order to produce a large number of them, mass-fabrication and mass-assembly by means of Microtechnologies should be pursued.

M. Cheng (Ed.): NanoNet 2008, LNICST 3, pp. 107–119, 2009.
© ICST Institute for Computer Sciences, Social Informatics and Telecommunications Engineering 2009

The cooperation between swarming agents is the key to the accomplishment of a desired task of the swarm. Communication plays a primary role in that, requiring important features to the communication system, without demanding too high resources. Communication capabilities in a microrobot are strongly limited by the microrobot size and power available on board. The first implies that only miniaturized communication systems can be integrated, while the latter imposes strict limits on communication distance and bit-rate. In addition, in order to exploit mass-fabrication, the system itself has to be relatively simple to allow automatic fabrication and assembly procedures. Following these constraints and a vast survey of possible communication technologies, optical communication demonstrated to be currently the only suitable and feasible solution for millimeter-sized microrobots. Indeed, also some of the most relevant multi-agent or swarm systems of inch-sized microrobots exploit (mainly) optics as a communication mean, e.g. iRobot [2], Alice [3] and Jasmine [4] robotic swarms.

Up to author's knowledge nobody has never attempted the mass-production of optical communication modules for millimeter-sized autonomous microrobots as those developed in the I-SWARM project.

1.1   The I-SWARM Project

The challenge to develop a miniaturized communication module to be integrated in (one of) the up-to-date world's smallest autonomous robot has been attempted in the frame of the I-SWARM project [5] [6], under the European Future and Emerging Technologies (IST-FET) Programme. The project aims to mass-produce autonomous millimeter-sized microrobots, which can then be employed as a "real" swarm capable to demonstrate observable emergent self-organization effects similar to those observed within ecological systems like ant states, bees colonies and other insect aggregations.

Only a few millimeter-sized autonomous robots (with limited functionalities) have been demonstrated in literature, e.g. [7], however, none of them approached the problem from a mass-production viewpoint. Actually, the concept of Swarm Microrobotics becomes (particularly) meaningful and powerful in correlation with this last issue.

The I-SWARM microrobots consist of a stack of assembled chip-modules for a whole size of about $3 \times 3 \times 3$ mm^3. An overview of the robotic modules and assembly process is in [8]. One of the final CAD design of the microrobot and a first prototype are reported in Fig. 1. Several tens of microrobots have already been assembled with an automatic machine-based process and are currently under testing. Each robot has a weight and volume of less than 70 mg and 23 mm^3 respectively. Although complete functionality of assembled robots is still heavily affected by the yield of the fabrication of each modules and of the assembly process itself, it is due to point out one of the most relevant results of the project from the hardware viewpoint: the establishment of a (preliminary) method for mass production and assembly of "chip-robots", a goal envisioned by some robotic experts in the past, e.g. [9], but a challenge never completely faced.

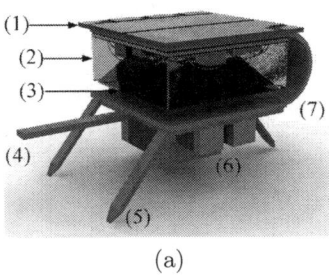

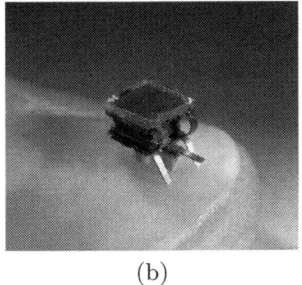

(a)   (b)

Fig. 1. CAD models of a final version of the I-SWARM autonomous microrobot; a) CAD drawing of the fully assembled robot (© P. Corradi (2007)): (1) Solar cell for energy scavenging from a double-lamp system equipping a custom small arena where microrobots will operate; (2) Optical communication module [10] [11]; (3) Electronics (Application Specific Integrated Circuit - ASIC) [12]; (4) Vibrating contact sensor: a vibrating cantilever with feedback sensor to be employed as touch-sensor to locate object/obstacles [13]; (5) Piezoelectric P(VDF-TrFE) legs, which are made vibrating for moving the robot [14]; (6) Capacitors; (7) Flexible printed circuit (FPC) backbone; b) One of the produced I-SWARM microrobots on a human thumb nail; this robot version is slightly different from the CAD model in (a) (courtesy of the I-SWARM Consortium and Uppsala University)

2 A Miniaturized Communication Module for Microrobots

Bio-inspiration has been the original approach in the effort to conceive a communication system that could let the development of swarm strategies demonstrated by some insects (e.g. ants, bees, wasps and termites). Therefore, initial work was focused on studying and trying to technologically conceive devices able to reproduce the interaction systems and methods that nature has evolved among insects of a same swarm. One of the most diffused communication systems used in natural swarms is based on the release of a chemical called *pheromone*. This demonstrated to be an extremely efficient technique in nature to develop emergent behaviours in swarms. However, the technical development of systems able to reliably release and detect chemicals along the time is a critical issue. As a consequence, several attempts to design communication systems for millimeter-sized robots were based on standard technological approach such as radio wave transmission or magnetic induction (e.g. [15]), however, only optics showed to be feasible according to the space and power constraints on board the microrobot (where the communication module has to share with the locomotion module a power budget of less than 0.8 mW at 3.6 V, and integrated capacitors can guarantee only very short and limited high power pulses for information transmission) and suitable for developing swarm strategies, mainly due to its feature of directionality in signal transmission.

In literature, there are some relevant examples of miniaturized communication systems (more conceived as nodes in communication networks rather than for equipping microrobots). In the frame of the *Smart Dust* project a device has been demonstrated based on a system of passive reflection of a LASER-based optical signal [16]. The system disadvantageously requires high voltages (about 100 V) for actuation. A similar miniaturized communication system has been introduced by the *Speckled Computing* project [17], but to the authors' knowledge not a single prototype has been built so far. Moreover, both the mentioned systems cannot assure proper information broadcasting in all the directions as required by swarm applications, because they are conceived more for far-range and focused optical emission. In addition, the conceived architecture is not likely to be easily produced by means of mass-fabrication processes, a basic requirement for the development of a swarm of microrobots.

2.1   Hardware Description

The optical module was completely designed taking into account mass-fabrication issues, and aiming at minimize the architecture, while reaching a sufficient functionality. The final design consisted of a 9 mm^3-body composed of a thin substrate, sub-mm sized photoemitter and photodetector dies (ELC-870 series and EPC-880-0.5 respectively, from Epigap, working in the $870 - 880 \, nm$ light range), which are integrated and wire-bonded along the borders of the substrate and controlled by the microrobot ASIC, 0201-sized surface-mounted resistors in the centre and a reflecting structure placed over the substrate, aimed at deflecting the incoming signals towards the photodetectors and at deflecting the signals generated and vertically emitted by the photoemitters toward the surroundings. The mirror structure was fabricated as a moulded transparent polymer body with a central pit with reflective 45°-sloped walls, covering and embedding the assembled devices. A CAD design is shown in Fig. 2(a). In Fig. 2(c) some of the final and fully functional fabricated optical modules: the polymer mirror looks black-colored because a visible light-blocking black-dye (Epolight^{TM} 7276A) was mixed into the polymer for ambient light rejection (up to about 850 nm). The pyramidal pit of the mirror were sputtered with chromium, in order to form reflective surfaces to deflect the light.

(a)	(b)	(c)

Fig. 2. a) Final CAD design of the optical module (size: $3 \times 3 \times 1.2 \, mm^3$), with LEDs (small boxes) and photodiodes (larger boxes) along the borders of the substrate, and resistors in the center; the modules has to be turned up-side down in the microrobot, see Fig. 1(a); (b) Mass-produced optical modules diced by LASER machine (mirrors are not sputtered); (c) Close view of mass-produced optical modules (one is turned up-side down showing the bottom-side electrical contacts) with metal sputtered mirrors.

2.2   Communication Properties

The optical properties of the communication module were firstly simulated before fabrication and experimentally characterized afterwards [11]. Tests were carried out to determine its radiation pattern and also its communication range, both in laboratory conditions and under the nominal illumination conditions of the I-SWARM arena. In Fig. 3 the set of measured points around the module with the same detected value of emitted light intensity, starting from a fixed value (e.g., the free-error communication intensity value at 15 mm distance) is reported and compared to the corresponding theoretical pattern reported as dotted line (and calculated starting from the consideration that the emission from each side of the optical module is Lambertian). The measured pattern results wider on the LED side due probably to bulk and surface scattering of the light in the polymer of the optical module. A significant overlapping of the free-error communication zone between adjacent sides of the module would evidently occur. This feature might be advantageous in order to improve the angular resolution in communication, by exploiting serial emission of the four LEDs (each identified with a specific code, see the next section 2.3) of the transmitting microrobot: in the signal-overlapping zone the receiving microrobot will be able to receive, consecutively in time, both the signals emitted by adjacent sides of the microrobot, thus better understanding the relative orientation of the transmitting microrobot.

For signaling, a defined digital protocol has been used [18], where only extremely short pulses (30 μs long) are emitted. Each single bit is started by a pulse; if this pulse is followed by a second pulse (after a software-adjustable time period), then the symbol is interpreted as a logic '1', if the second pulse is missing,

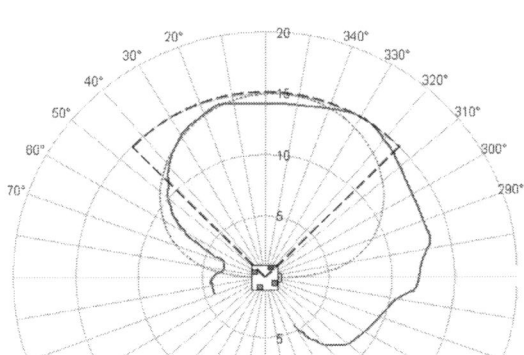

Fig. 3. Polar plots of the error-free communication pattern generated by one side only of the optical module: the continuous line plots measured data starting from a maximum distance of 15 mm; the dotted elliptical-like line represents the calculated radiation pattern; the sectioned darker line, defining a 90°-wide circular sector, is the radiation pattern as theoretically considered in the algorithms presented in the following and as it is modeled in the simulation described in the section 3. The module size in the centre is exaggerated for clarity (it should be only a central point). Units are degrees for the angles and mm for the distance.

as a logic '0'. A beginning of a data frame is marked by three consecutive pulses. The ASIC hardware supports a frame length between 1 and 32 bits. Bit rate can be adjusted by software from 83 bps to 2083 bps. The average maximum power consumption on a 32-bit frame at the lowest transmission speed is $36.6\mu W$ (438.5 μW at 1 $Kbps$), with a peak power consumption of 7.2 mW (2 mA at 3.6 V). Error-free directional communication capability was demonstrated with the optical modules up to 20 mm in standard laboratory conditions with a supplied current of 2 mA and using standard laboratory electronics. By using the ASIC both for transmission and reception the same distance decreased down to 9 mm. Due to powering illumination in the final arena set-up, the inter-robot communication distance was further reduced to 4.5 mm. Several realistic improvements to increase performances of the whole system have been considered for future works.

2.3   A Basic Communication Strategy between Microrobots

The introduced optical system allows a basic directional communication strategy for both collision avoidance between microrobots and a cooperative behavior without any external supervision, as described in the following. A similar technique is described in [19] for multi-agent robotic applications, but for much larger robots. During communication each LED belonging to one microrobot is identified with a particular bit string; in the case of four LEDs, two-bit strings are enough: 00, 01, 10, 11 (Fig. 4(a)). In this way surrounding microrobots can detect not only the presence and position of one or more microrobots, but also understand if it/they are on a direct collision course and react accordingly. The mentioned cooperative strategy is illustrated in Fig. 4. The strategy is based on the following steps: a microrobot MR_1 finds a target or obstacle and a microrobot MR_2 enters MR_1's communication range. In a more general configuration, the microrobot MR_{N+1} needs to know:

A. The relative direction of the microrobot MR_N (which is understood depending on which of the MR_{N+1}'s photodiode(s) receives the signal);
B. The relative orientation of MR_N (received by bit communication);
C. The target direction vector with reference to MR_N (received by bit communication).

Combination of points B and C transposes the vector of the target from the reference of MR_N to the reference of MR_{N+1}. In this way the vector named $\boldsymbol{V}$ in Fig. 4(b) is acquired from MR_{N+1} relatively to its own reference system. The final vector $\boldsymbol{V_F}$ for the target direction is obtained by calculating the vector sum of the components of the acquired vector $\boldsymbol{V}$ and the vector $\boldsymbol{V_R}$, introduced in point A. This allows each microrobot to know the direction to the target (and eventually also the distance, if $\boldsymbol{V_R}$ is thought to have a known value, proportional to, for instance, the detected intensity of the received signal). Fig. 4(c) shows an example of the propagation of this strategy to several members of the swarm and the formation of a vector trail. In this basic form, it is possible to think that as soon as the microrobot MR_{N+1} enters the MR_N's communication range, it stops

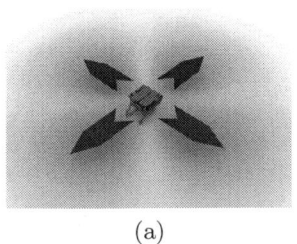 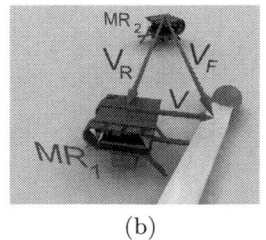 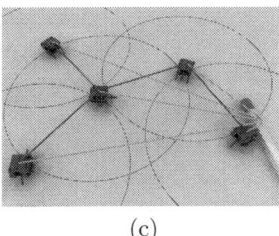

(a)   (b)   (c)

Fig. 4. (a) Representation of a theoretical emission pattern of the optical module mounted into the microrobot; (b) Vector sum for reconstruction of the direction of the target (e.g. a match): $V_F = V_R + V$; (c) Optical network established inside the swarm: darker vectors are the V_R for each microrobot and define the forming trail; the circles defines the theoretical border of the communication range for each microrobot. © P. Corradi (2007)

and starts to forward the signal. In this way the technique allows information broadcasting within the swarm, creating a motionless communication network, that extends in time, thus increasing progressively the probability that other microrobots could meet it and, therefore, receive information.

The main issue regarding the working principle of the developed communication module and the introduced basic communication strategy consists in a poor angular resolution for the identification of direction and relative orientation between microrobots, because only four directions basically can be discriminated. Nevertheless, the bio-inspired swarm algorithm, presented in the next section 3, which is based on these basic rules, demonstrates in simulations collective swarm behaviours and the establishment of an emergent network, which, in this case, is also "mobile" (microrobots keep moving while receiving messages). Although the optical properties of the described communication module are modeled in the simulation in a simplified way (in particular, as clearly visible in Fig. 3, the modeled radiation pattern does not reproduce a considerable part of the measured optical pattern), the results are significant because they introduce a preliminary demonstration of the possibility to implement swarm strategies on the base of the minimal communication module developed and the simple rules described. The same strategies will be tested in the final physical microrobotic swarm as soon as enough fully functional microrobots will be available.

3   A Bio-Inspired Vector-Based Swarm Algorithm

For swarm-robotic algorithms, nature offers a variety of sources of inspiration by providing a multitude of biological solutions to "swarm problems". It is desired to find algorithms that are easy enough to be implemented with the limited computational capabilities available on board autonomous swarm microrobots. Besides that, the algorithms should be robust enough to work in a noisy environment, sensed through imperfect sensors, flexible enough to be able to deal with rapidly changing environments. Obviously, the algorithms should also scale well to be used in ever-increasing swarm sizes, as they are used in today's swarm robotics [20]. These constraints require a decentralized local-neighbour based communication, as it is frequently found in natural swarm systems [21]. On the base of the features of the introduced communication module and on the described vector-transmission technique, we describe here the basic principles of a bio-inspired communication and navigation algorithm, which is based on simple vector communication, as it is found in honeybees [22], and on vector summation, as it is found in the desert ant Cataglyphis [23] [24]. We demonstrate here, how this vector-based algorithm can allow hundreds of robots to allocate themselves at the right places in the arena and to organize themselves into self-organized trails in a multi-source/one-target scenario.

The algorithm is based on the communication of simple vectors within a swarm of microrobots equipped with the described communication module. This was modeled in the simulations merely at a functional level, characterized by a set of 4 light emitters and sensors, which emit and perceive horizontally in four directions (front, rear, left, right) with 90° between the central transects of each emitted light beam. No light physics was implemented at this stage, the simulation limited to show the emergent cooperative effect based on the theoretical working principle of the optical module, although several realistic conditions were implemented, see in the following. The communicated messages consist of 3 integer values and 3 boolean signals: The Boolean (On/Off) signals indicate the internal status of the sending robot, two integer values are used for communicating the vector towards the target, and one integer value is used as a hop-count, which allows to identify the "age" of the message. The necessary calculations performed within the robots are simple additions of vectors as well as "if-else" statements.

The testbed for the proposed algorithm is the simulator "LaRoSim v.66" [25], which is a multi-agent bottom-up simulation of a swarm of I-SWARM robots. The robotic swarm is tested in a cleaning scenario, in which the robots have to encounter dirt areas (sources), attract other robots to these source areas and then move on an as-short-as-possible path to a designated dump area (target). The robots can sense the source area and the target area only if they are already located there and can then communicate messages to other nearby robots within a circular neighbourhood of 3.5 robot-diameters. The vector-based algorithm is described as follows:

1. All robots start unloaded (no dirt particles loaded) at randomized positions in the arena, are headed in randomized directions, and move straight ahead.
2. If a robot senses another robot or an obstacle in front, it turns away of the encountered obstacle.
3. If an unloaded robot realizes that it is located on a source area, it picks up a particle and emits a specific boolean signal (*signal-1*) and an additional signal coding for the LED used. An additional hop-count is also attached to each message. As this robot is located directly on the source area, the hop-count is set to 1, what indicates information of the highest possible quality. After some time-steps, this robot picks up a dirt particle and significantly

changes its internal status this way. See step 7 for this robot's further behavioural rules.

4. If another robot receives the *signal-1* and the corresponding orientation code, it calculates the relative angle to the robot that is located on the source area and adds the vector (V) that was transmitted by communication. This way, the receiving robot can calculate the resulting vector (V_F), which should point directly to the robot located on the source. The robot then starts to emit another specific boolean signal (*signal-2*), which indicates that the robot is not located on the source area, but is receiving "high-quality" information directly from a robot that is located on the source area. It also emits the vector towards the source area, as well as a beam-specific (LED) code. At the end of the message, a hop count (now increased to 2) is sent.

5. Other robots can receive this message and can update their own vector towards the source area, as long as the hop-count of the received message is below or equal to the already stored information.

6. All unloaded robots that receive such a vector message turn towards the location of the source area and move a small distance forward before they receive new information, calculate new resulting vector by vector summation and they transmit this new vector again to their neighbours.

7. For describing the behaviour of loaded robots, the same rules as mentioned in the steps 3-6 apply, except that the term "unloaded" has to be replaced by the term "loaded" and the term "source" has to replaced by the term "target": These robots move in trail formation from the source areas (dirt) towards the target areas (dumps). As soon as they reach the dump area, they drop the carried dirt particle there, change their status to "unloaded" and continue with step 1.

For testing this algorithm in the simulator under realistic conditions, an error in communication (*P_break_communication*) was assumed. Furthermore we assumed that robots cannot measure distances and could not perform angular measurements to other robots, they can only discriminate the side the other robot is located relative to themselves (front, rear, left, right) by evaluating the photodetector that received the message from the other robots. For calculating the vectors to other robots, the robots use a "standard distance", which reflects the half of the maximum communication range, and they use a "standard angle", which reflects the median transect of the area covered by the receiving photodetector. To allow outdated (not-reinforced) information to leave the system, two additional rules were implemented: With a given probability of 20%, a robot spontaneously forgets its information and refreshes its internal memory by accepting information from other nearby robots. Except for this case, robots accept only communicated vectors with hop-counts less or equal to their own stored "old" information. To prevent the robots from aggregating too densely in specific parts of the arena (e.g. source areas, target areas), we prevented a fixed fraction (10%) of the robots from navigating towards the points that are communicated by the vectors. These robots ("scouts") perform just the random walk and communicate with neighbours, thus their important role is to provide a bridge in the swarm network from one aggregation area to the others.

3.1   Results

The simulated arena setup in the cleaning scenario consists of an arena wall (outer boundary), four dirt areas (sources) in the corners of the rectangular arena, and one central target area (dump), see Fig. 5.

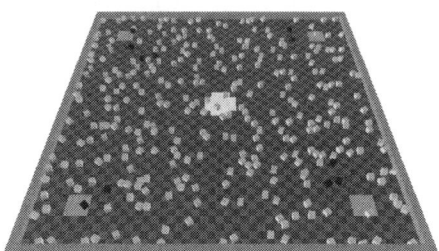

Fig. 5. Screenshot of the simulation scenario: Light-grey boxes indicate unloaded robots. Dark-black boxes indicate loaded robots. They move on a chessboard-like arena, where every grey square represents a patch. In the corners of the arena, four rectangular dirt areas (sources) are shown. In the center of the arena, a cross-shaped dump area (target) is shown. Robots should transport dirt particles from the sources to the target in as direct as possible trails.

By varying the parameter *P_break_communication* and the number of robots on the arena (i.e. the swarm density in the arena, expressed in percentage by the parameter *swarm_density*), we show the robustness of the algorithm. To visualize the results, we tracked the paths of all loaded robots and coded them as shades of grey on the arena floor. The darker an area is, the more loaded robots have been located on that patch of the arena. As Fig. 6 shows, both parameters affect significantly the directness of the robots' motion. However, under all tested circumstances the robot swarm was always able to form a trail heading clearly towards the central dump area.

As the probability of communication breaks increases (Fig. 6, from left to right column), the trails get (slightly) wider, indicating less optimal navigation towards the central target area. As the swarm intensity increases (Fig. 6, from top to bottom row), the global swarm behaviour changes: With a density of 5% (*swarm_density* = 0.05, corresponding to a swarm size of 109 robots, Fig. 6(a),(b),(c)), no trails emerge at all. The robots' motion show just random trajectories, because the distances between robots are too large to allow longer chains of robot-to-robot communication, thus no network is estalished. With a swarm density of 15% (i.e. 328 robots, Fig. 6(d),(e),(f)), the robots clearly approach the central target. The smaller the communication failure rate, the clearer the emerging trails are (compare Fig. 6(d),(e),(f)). With a *swarm_density* of 25% (i.e. 547 robots, Fig. 6(g),(h),(i)), still prominent trails emerge. But most robots accumulate in a ring-shaped structure around the target, not on the target area itself. The high swarm density leads to the fact that a large number of robots that were initially located on the target area got trapped by the dense trail

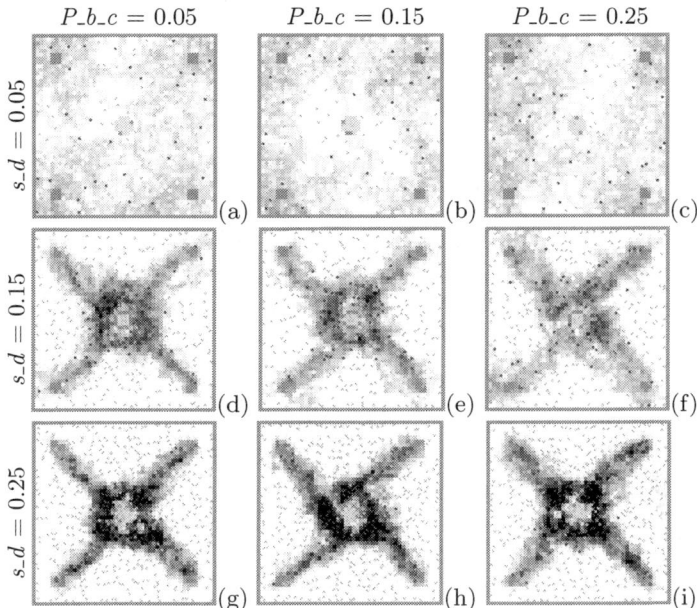

Fig. 6. The effect of the parameters *swarm_density* (*s_d*) and *P_break_communication* (*P_b_c*) on the directness of the swarm's navigation

heading towards this target area. Thus, a robot density of 15% was found to be a near-optimal swarm-density for the given scenario.

Finally we wanted to investigate how these parameters affect the efficiency of the robotic swarm. For all the 9 parameter combinations, corresponding to Fig. 6(a)-(i), the number of picked-up and delivered dirt particles was measured. The result states that a swarm density of 15% and a value of *P_break_communication* of 5% is optimal for particle delivery. A higher swarm density can increase the number of picked-up dirt particles but results in a significantly lowered number of delivered particles due to the emergence of the dense ring-shaped robot trail around the delivery area.

4 Conclusions

We have introduced a novel mass-producible miniaturized $3 \times 3 \times 1.2$ mm^3 communication system for swarming microrobots, whose architecture allows the exploitation of a simple communication strategy based on the transmission of vectorial information. The possibility to form communication networks between more and more tiny mobile robots is indeed based on solutions of minimal, thought functional, communication hardware and the exploitation of distributed and decentralized intelligence in suitable large multi-agents systems or swarms. The presented bio-inspired vector-based algorithm is a robust and computational easy algorithm, which bases on the working principle of the developed optical module and requires only very limited computational power of the robots it is executed on. With 3 Boolean and 3 integer values (per motion cycle), the required bandwidth of communication is low (less than 17 *bps*). Simulation results show that using simple navigation rules and limited nearest-neighbour communication, a desired and well directed collective swarm behaviour, e.g., trail formation, can be achieved. The robotic swarm demonstrated to work with a variety of parametrizations, as well as with a high number of target areas, leading to a complex pattern formation of autonomously emerging trails of loaded robots all heading towards a single target, starting from several "trail sources".

Future works include modeling the measured emission/reception radiation pattern of the fabricated optical module in the simulator, and, finally, experimental tests with the fabricated microrobots.

Acknowledgments. The work described has been funded by the I-SWARM project (grant IST-2004-507006) of the Future and Emerging Technologies (IST-FET) Programme within the European 6th Framework Programme.

References

1. Bonabeau, E., Dorigo, M., Theraulaz, G.: Swarm intelligence: from natural to artificial systems. Oxford University Press, New York (1999)
2. McLurkin, J., Yamins, D.: Dynamic Task Assignment in Robot Swarms. In: Robotics Science and Systems Conference, Cambridge, MA, USA (2005)
3. Caprari, G., Balmer, P., Piguet, R., Siegwart, R.: The Autonomous Micro Robot "Alice": a platform for scientific and commercial applications. In: International Symposium on Micromechatronics and Human Science, Nagoya, Japan (1998)
4. Kornienko, S., Kornienko, O., Levi, P.: Collective AI: context awareness via communication. In: 19th International Joint Conference on Artificial Intelligence, Edinburgh, Scotland, pp. 1464–1470 (2005)
5. I-Swarm (Intelligent-Small World Autonomous Robots for Micromanipulation), 6th Framework Programme Project No. FP6-2002-IST-1. European Community (2003–2007), http://www.i-swarm.org
6. Seyfried, J., Bender, N., Estaña, R., Szymanski, M., Thiel, M., Woern, H.: Design and scenarios for a real nano-manipulation robot swarm. In: Mechatronics and Robotics, Aachen, Germany, pp. 1368–1373 (2004)
7. Hollar, S., Flynn, A., Bellew, C., Pister, K.S.J.: Solar powered 10 mg silicon robot. In: IEEE International Conference on Micro Electro Mechanical Systems, Kyoto, Japan, pp. 706–711 (2003)
8. Edqvist, E., Snis, N., Gao, J., Scholz, O., Casanova, R., Diéguez, A., Corradi, P., Wyrsch, N., Johansson, S.: Flexible building technology for microsystems: Surface assembly of a mass produced millimeter sized microrobot. J. of Micromechanical Systems (submitted) (2008)
9. Flynn, A.M.: Gnat Robots (And How They Will Change Robotics). In: IEEE Micro Robots and Teleoperators Workshop, Hyannis, MA, USA (1987)
10. Corradi, P., Ranzani, L., Scholz, O., Diéguez, A., Menciassi, A., Laschi, C., Martinelli, M., Dario, P.: Free-Space Optical Communication in a Swarm of Microrobots. In: European Conference on Optical Communications, Berlin, Germany (2007)

11. Corradi, P., Scholz, O., Knoll, T., Menciassi, A., Dario, P.: Optical System for Communication and Sensing in Millimeter Sized Swarming Microrobots. J. of Micromechanics and Microengineering (submitted) (2008)
12. Casanova, R., Dieguez, A., Sanuy, A., Arbat, A., Alonso, O., Canals, J., Samitier, J.: An ultra low power IC for an autonomous mm^3-sized microrobot. In: International Solid-State Circuits Conference, San Francisco, CA, USA, pp. 55–58 (2007)
13. Corradi, P.: et al.: Study of a multifunctional vibrating micro-cantilever for application in Microrobotics (unpublished)
14. Edqvist, E., Snis, N., Johansson, S.: Gentle Dry Etching of P(VDF-TrFE) Mulilayer Micro Actuator Structures by use of an ICP. J. Micromech. Microeng. 18(1) (2008)
15. Kim, S., Knoll, T., Scholz, O.: Feasibility of Inductive Communication Between Millimeter-Sized Wireless Robots. IEEE Trans. on Robotics 23(3), 605–609 (2007)
16. Zhou, L., Kahn, J.M., Pister, K.S.J.: Corner-Cube Retroreflectors Based on Structure-Assisted Assembly for Free-Space Optical Communication. J. of Microelectromechanical Systems 12(3) (2003)
17. Arvind, D.K.: Speckled Computing. Invited talk IEEE Colloquium Series, Edinburgh, Scotland (2005)
18. Alonso, O., Casanova, R., Arbat, A., Sanuy, A., Diéguez, A., Scholz, O., Corradi, P., Samitier, J.: An Optical Interface for Inter-Robot Communication in a Swarm of Microrobots. In: 1st International Conference on Robot Communication and Coordination, Athens, Greece (2007)
19. Payton, D., Daily, M., Estkowski, R., Howard, M., Lee, C.: Pheromone robots. Autonomous Robots 11, 319–324 (2001)
20. Sahin, E.: Swarm Robotics: From Sources of Inspiration to Domains of Application. In: Şahin, E., Spears, W.M. (eds.) Swarm Robotics 2004. LNCS, vol. 3342, pp. 10–20. Springer, Heidelberg (2005)
21. Camazine, S., Deneubourg, J.L., Franks, N., Sneyd, J., Theraulaz, G., Bonabeau, E.: Self-Organization in Biological Systems. Princeton University Press, New York (2001)
22. Frisch, K.V.: Tanzsprache und Orientierung der Bienen. Springer, Berlin (1965)
23. Andel, D., Wehner, R.: Path integration in desert ants, Cataglyphis: how to make a homing ant run away from home. Royal Society London B 271, 1485–1489 (2004)
24. Collett, M., Collett, T.S., Bisch, S., Wehner, R.: Local and global vectors in desert ant navigation. Nature 394, 269–272 (1998)
25. Schmickl, T., Crailsheim, K.: Trophallaxis within a robotic swarm: bio-inspired communication among robots in a swarm. Autonomous Robots 25, 171–188 (2008)

Counting Photons Using a Nanonetwork of Superconducting Wires

Andrea Fiore[1], Francesco Marsili[1,2], David Bitauld[1], Alessandro Gaggero[3],
Roberto Leoni[3], Francesco Mattioli[3], Aleksander Divochiy[4], Alexander Korneev[4],
Vitaliy Seleznev[4], Nataliya Kaurova[4], Olga Minaeva[4], and Gregory Gol'tsman[4]

[1] Eindhoven University of Technology, P.O. Box 513, NL-5600MB Eindhoven,
The Netherlands
a.fiore@tue.nl
[2] Ecole Polytechnique Fédérale de Lausanne (EPFL),
Institute of Photonics and Quantum Electronics (IPEQ), Station 3,
CH-1015 Lausanne, Switzerland
[3] Istituto di Fotonica e Nanotecnologie (IFN), CNR, via Cineto Romano 42,
00156 Roma, Italy
[4] Moscow State Pedagogical University (MSPU), Department of Physics,
119992 Moscow, Russian Federation

Abstract. We show how the parallel connection of photo-sensitive superconducting nanowires can be used to count the number of photons in an optical
pulse, down to the single-photon level. Using this principle we demonstrate
photon-number resolving detectors with unprecedented sensitivity and speed at
telecommunication wavelengths.

A superconducting nanowire biased close to the critical current can be used as a
single-photon detector [1]: As a photon is absorbed, a nanoscale hot-spot is formed,
producing a resistive transition across the wire, which is detected as a voltage pulse in
the external circuit. However, a single wire (usually patterned into a meander shape to
increase the active area) has a response nearly independent of the number of incident
photons, as the resistance produced by the absorption of a single photon is sufficient
to divert all the current to the external circuit. We have recently proposed [2] a novel
detector architecture, the "Parallel Nanowire Detector" (PND), which provides an
output electrical pulse whose amplitude is proportional to the incident photon number.
The basic structure of the PND is the parallel connection of N superconducting
nanowires, each connected in series to a resistor R_0 (Fig. 1). In this parallel
configuration, the currents from different wires can sum up on the external load,
producing an output voltage pulse proportional to the number of photons.

PNDs were fabricated on ultrathin NbN films (4nm) on MgO and R-plane sapphire
using electron beam lithography (EBL) and reactive ion etching. Detector size ranges
from 5x5 μm^2 to 10x10 μm^2 with the number of parallel branches varying from 4 to 14.
The nanowires are 100 to 120 nm wide and the fill factor of the meander is 40 to 60%.

M. Cheng (Ed.): NanoNet 2008, LNICST 3, pp. 120–122, 2009.

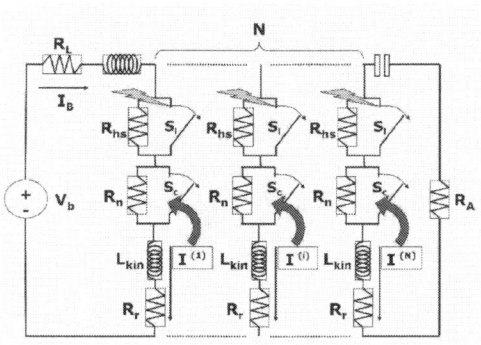

Fig. 1. Equivalent circuit of a PND

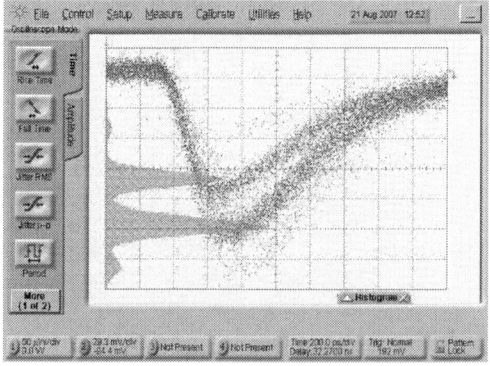

Fig. 2. Oscilloscope histograms during photodetection by a PND with 4 parallel wires. Up to four photons are detected.

The length of each nanowire ranges from 25 to 100 μm. Designs with and without the integrated bias resistors were tested.

The photoresponse of a PND with four parallel wires probed with light at 1.3 μm was recorded by a sampling oscilloscope (Fig. 2). All four possible amplitudes can be observed. The pulses show a full width at half maximum (FWHM) as low as 700 ps. PNDs showed counting performance when probed with light at 26 and 80 MHz repetition rate, outperforming any existing PNR detector at telecom wavelength by three orders of magnitude.

The one-photon quantum efficiency η at 1.3 μm and dark-counts rate DK were measured as a function of bias current. The lowest DK value measured was 0.15 Hz for $\eta \sim 2\%$ (yielding a noise equivalent power $NEP=5.6\times10^{-18}$ W/Hz$^{1/2}$), limited only by the room temperature background radiation coupling to the PND. Additionally, unlike most PNR detectors, no multiplication noise can be observed in PNDs, as the width of the histogram peaks is independent of the number of detected photons.

In conclusion, a new photon-number-resolving detector, the Parallel Nanowire Detector, has been demonstrated, which significantly outperforms existing approaches in terms of sensitivity, speed and multiplication noise in the telecommunication wavelength range. The ability to measure the photon number is a key asset in quantum optical information processing, where states with a well-defined photon number are routinely used for the transmission and processing of quantum information. Additionally, by further extending this concept we envisage the fabrication of analog detectors with large dynamic range (>30 photons) and single-photon sensitivity, which would bridge the gap between conventional detectors and single-photon detectors, for applications in optical sensing and communications.

References

1. Gol'tsman, G.N., Okunev, O., Chulkova, G., Lipatov, A., Semenov, A., Smirnov, K., Voronov, B., Dzardanov, A., Williams, C., Sobolewski, R.: Applied Physics Letters 79(6), 705 (2001)
2. Divochiy, A., Marsili, F., Bitauld, D., Gaggero, A., Leoni, R., Mattioli, F., Korneev, A., Seleznev, V., Kaurova, N., Gol'tsman, G., Lagoudakis, K., Benkhaoul, M., Lévy, F., Fiore, A.: Nature Photonics 2, 302 (2008)

Communicating Mobile Nano-Machines and Their Computational Power⋆

(Extended abstract)

Jiří Wiedermann[1] and Lukáš Petrů[2]

[1] Institute of Computer Science, Academy of Sciences of the Czech Republic,
Pod Vodárenskou věží 2, 182 07 Prague 8, Czech Republic
jiri.wiedermann@cs.cas.cz
[2] Faculty of Mathematics and Physics, Charles University,
Malostranské náměstí 25, 118 00 Prague 1
Czech Republic
lukas.petru@st.cuni.cz

Abstract. A computational model of molecularly communicating mobile nanomachines is defined. Nanomachines are modeled by a variant of finite-state automata—so-called timed probabilistic automata—augmented by a severely restricted communication mechanism capturing the main features of molecular communication. We show that for molecular communication among such motile machines an asynchronous stochastic protocol originally designed for wireless (radio) communication in so-called amorphous computers with static computational units can also be used. We design an algorithm that using the previous protocol, randomness and timing delays selects with a high probability a leader from among sets of anonymous candidates. This enables a probabilistic simulation of one of the simplest known model of a programmable computer—so-called counter automaton—proving that networks of mobile nanomachines possess universal computing power.

Keywords: molecular communication; nanomachines; timed probabilistic automata; communication protocol; universal computing.

1 Introduction

Nanomachines are molecular cell-sized artificial devices or engineered organisms produced by self-assembly or self-replication, capable of performing simple tasks such as actuation and sensing (cf. [4]). Construction of various nanomachines seems to be entirely within reach of current nano- and bio-technologies (cf. [3]). Once constructed and endowed with a certain sensor and actuating abilities communication among nanomachines becomes an important problem.

⋆ This research was carried out within the institutional research plan AV0Z10300504 and partially supported by the GA ČR grant No. 1ET100300419 and GD201/05/H014.

M. Cheng (Ed.): NanoNet 2008, LNICST 3, pp. 123–130, 2009.

Corresponding molecular mechanism design presents a great challenge for nano-technology, bio-technology, and computer science. In molecular biology there is an explosively growing field dealing with molecular communication. Here, the respective research is interested almost exclusively in the (bio)chemical aspects of existing communication mechanisms in living systems. It seems that almost no attention has been paid to the algorithmic aspects of the respective communication process. Communication at a nanoscale substantially differs from communication scenarios and frameworks known from classical distributed systems. Key features of traditional versus molecular communication have been neatly summarized in [4]. In molecular communication, the communication carrier is a molecule; chemical signals are extremely slowly propagated by diffusion in an aqueous environment with low energy consumption. In general, it is not necessary that the signal will reach all targets: a majority will do. This is to be compared with traditional communication via electromagnetic waves where electronic or optical signals are propagated at light speed with high energy costs in an airborne medium and message delivery to all targets is required.

In what follows we will concentrate onto a scenario in which nanomachines form an autonomous system operating in a closed liquid environment without external control (a similar scenario is also considered in [4]). The system consists of a finite number of nanomachines freely "floating" in their environment that interact via molecular communication creating thus a kind of ad-hoc network. We assume that there is a sufficient number of nanomachines such that within the "communication radius" (to be explained later) of each machine there are other nanomachines available. Within the communication process the authors of [4] have identified following steps: encoding of information onto the information molecules, sending of the carrier/information molecules into the environment, propagation of the carrier/information molecules through the environment, receiving of the carrier/information molecules, and decoding of the information represented by the received information molecules into reaction at a receiver. In addition, recycling of carrier/information molecules may be necessary to avoid accumulation at a receiver.

This seems to be a reasonable sequencing of communication subtasks, but from an algorithmic viewpoint several questions immediately arise. What protocol is used for molecular communication? What happens if there are more nanomachines communicating concurrently? Would it be necessary to synchronize them so that one would act as a sender and the other as a receiver? How does the sending machine learn that a target machine has received its signal? If we allow a finite number of different signals (types of molecules) by which the machines can communicate, what happens if a machine having several receptors detects different signals at different receptors at the same time? What happens when a (would be) receiving machine is engaged in sending a signal? What are the computational limits of the underlying system?

It is the purpose of the present paper to answer such questions. In order to do so three things are needed: (i) a more detailed algorithmic (computational) model of a nanomachine, (ii) a communication protocol, i.e., an algorithm controlling the communication behavior of each nanomachine, and (iii) a simulation procedure showing the relation of networks of communicating nanomachines to a standard (classical) model of computations whose computational power is known.

The contribution of the paper corresponds to the latter mentioned three items. First, the paper defines a new model of communicating nanomachines based on probabilistic timed finite state automata. This seems to be a novel application of such types of automata. The size of the resulting network of nanomachines is scalable even though the number of states of each nanomachine remains fixed (i.e., independent of the number of communicating machines). The most important feature of our nanomachine model is its part respecting the limitations of molecular communication. The communication mechanism works with the minimal functionalities available at the molecular level: a finite number of states, randomness, asynchronicity, anonymity of nanomachines, and one-way communication without a possibility of signal reception acknowledgement. Second, the paper shows that for basic communication among the mobile nanomachines a protocol originally designed for wireless radio communication in the case of so-called amorphous computers with the static computational units can be used. This points to the robustness of the respective protocol that is used in a completely different communication medium and for mobile, rather than static communicating units. Last but not least, we show that nanomachines captured by our modelling are able to perform whatever kind of computation—they possess universal computational power.

The structure of this extended abstract also mirrors the previous three items. In Section 2 the nanomachine computational model is introduced and in Section 3 it is shown that wireless communication protocol designed in [7] can be also used for the case of molecular communication. In Section 4 a simulation of a very simple model of universal computer—so-called counter automaton—by nanomachines is shown.

The full version of the paper is available as a technical report [8].

2 Nanomachine Computational Model

From computational point of view we will see each nanomachine as a timed probabilistic finite-state automaton. In essence this is a finite state automaton augmented with quantitative information regarding the likelihood that transitions occur and the time at which they do so. Timing is controlled by a finite set of real valued clocks. The clocks can be reset to 0 (independently of each other) with the transitions of the automaton, and keep track of the time elapsed since the last reset. The transitions of the automaton put certain constraints on the clocks values: a transition may be taken only if the current values of the clocks satisfy the associated constraints (cf. [2], [5]).

A biomolecular realization of a probabilistic automaton has been described in [1]. As far as timing mechanisms of biological automata are concerned, in biology there is a vast body of research dealing with biological oscillators and clocks controlling various biological processes in living bodies and it is quite plausible that such mechanisms could also be considered in nanosystems.

Our version of probabilistic timed automaton will in fact be a probabilistic timed transducer (Mealy automaton) having a finite number of input ports (receptors) and output ports (emitters). The signal molecules will be represented by elements of automaton's finite working alphabet. The conditions under which the nanomachines can communicate are designed so as to make the underlying molecular mechanism as simple as possible while capturing the constraints imposed on molecular communication:

1. Each automaton is able to work in two modes: in the receiving mode, reading (in parallel) the symbols (molecules) from its input ports, and in the sending mode, writing the same symbols to its output ports. A read operation is successful if and only if all symbols at all input ports are identical. Otherwise, when the symbols at the input ports differ, a so-called *communication collision* occurs: the read operation fails and the symbols are released from the input ports. Except of communication ports, a nanomachine can also have receptors detecting other than signaling molecule stimuli, and other actuators doing some job corresponding to the purpose to which the nanomachine has been designed.

2. In the sending mode, an automaton releases signal molecules of the same kind through all its output ports; these molecules diffuse in the environment and eventually can reach the input ports of an automaton in a receiving mode.

3. After a certain time, signal molecules disintegrate into other molecules which are not interpreted by the automata as signal molecules. In their life time, signal molecules can travel, by diffusion, in average a certain maximal distance called *communication radius*.

4. The automata work asynchronously — there is no global clock in the system. The actions of each automaton are governed by automaton's local clock. The local clocks in the automata are not synchronized, however, they all "tick" (roughly) at the same rate since they all are realized by the same biochemical oscillators. A slight variation (that in practice may also be caused, e.g., by temperature variations) in the clock rate does not harm our purposes.

5. The automata have no identifiers, i.e., for communication purposes all senders and receivers "look the same".

6. The automata are equipped by a finite set of timers (clocks) that are assigned to certain transitions. These timed transitions work as described in the beginning of the section.

Note that the previous conditions are quite restrictive. Condition 1 means that an automaton cannot simultaneously be in a sending and receiving mode; a signalling molecule reaching an input port of an automaton in a sending mode will not be detected by that automaton by that time. Condition 2 essentially says that if a broadcast from one automaton is "jammed" by a broadcast of an other automaton broadcasting different signal, then the receiving automaton does not accept any signal. Condition 3 ensures that signal molecules cannot "roam" for ever in the environment and that current signals eventually prevail over the old ones in the communication radii of the automata. Condition 4 says that we cannot count on automata switching their sending and receiving modes synchronously. Also note that our automata can move in their environment, either passively, due to the external forces (e.g., in a bloodstream), or actively, like some bacteria. This by itself, but the more so in conjunction with condition 5, prevents whatever kind of "acknowledgements" of received messages. Last but not least, observe that making use of a finite number of states (independent of the number of communicating automata) does not allow "storing" automata "addresses" (names) in automaton's states since the number of such addresses grow unboundedly with the number of nanomachines.

A multiset of communicating identical nanomachines is called a *nanonet*.

3   Communication Protocol

Thanks to a similarity between the computational model of "static" amorphous computer from [7] and the present model a similar communication protocol as in the latter case applies. In order to enable an algorithmic insight into the functionality of the protocol we briefly describe the necessary background.

Consider the so-called *communication graph* G of a given nanonet whose nodes are nanomachines (automata) and edges connect those automata which are within the communication radius of each other. The size of G will be measured in the number n of its nodes. Obviously, the topology of G depends on time since in general our automata move. In order to enable communication among all (or at least: a majority of) available automata in a nanonet most of the time G must have certain desirable properties. What we need are connected graphs with small diameter and a reasonable node degree. The most important property of G is its connectivity: connectivity is a necessary condition in order to be able to harness all processors. Graph diameter $D = D(n)$ bounds the length of the longest communication path. Finally, the node degree Q (i.e., the maximal neighborhood size of a node) determines the collision probability of signals.

In order to our communication algorithms work in the way we assume we will consider the families of so-called *well-behaved nanonets*. For any n, these are the nanonets whose underlying computational graph of size n stays connected and whose diameter and maximal neighborhood size stay bounded by $D = D(n)$ and a constant Q, respectively, all the time.

The requirements put on the well-behaved nanonets are quite strong and one can hardly imagine that in "practice" they will be fulfilled, indeed. Nevertheless, the invariance of these properties is needed in order to be able to analyze the correctness and efficiency of the communication primitives we will deal with in a sequel. After presenting these primitives we will see that they are sufficiently "robust" in order to operate correctly and with a similar efficiency also in instances of nanonets that occasionally, for short time, deviate from their well-behaved properties.

Protocol Send. For delivering a signal s from a node X to any node Y in its communication neighborhood with a given probability $\varepsilon > 0$ of failure a wireless *Protocol Send* designed in [7] is used.

The idea is for each node to broadcast sporadically, minimizing thus a communication collision probability in one node's neighborhood. This is realized as follows. Let T be time to transfer a signal between any two neighbors at the communication radius distance. Each node has a timer measuring *timeslots* (intervals) of length $2T$. During its own timeslot, each node is allowed either to listen till the end of its timeslot, or to send a signal at the very beginning of its timeslot and then listen till the end of this timeslot. At the start of each timeslot a node sends s with probability $p = 1/(Q+1)$ and this is repeated for $k = O(Q\log(1/\varepsilon))$ subsequent timeslots. The probability of a node sending s is controlled by the transition probability of the respective probabilistic automaton.

Assuming that all nodes send their signals asynchronously according to *Protocol Send,* in [7] it is shown that s will be received by Y in time $O(Q\log(1/\varepsilon))$ with probability at least $1 - \varepsilon$.

In order to allow the sending automaton to send s $k = k(\varepsilon)$ times in a row as required by the protocol its timer must be set to the interval $\approx kT$. Note that this is the time for which a node must be in the sending mode.

Algorithm Broadcast. In order to send a signal from a node to any other node of a nanonet which is not in the communication radius of the sending node the idea of flooding the network by that signal is used.

The main idea of *Algorithm Broadcast* is to use every node reached by a given signal s as a "retranslation station" that distributes s using *Protocol Send* further through the network. After retransmitting s each node stops retransmitting s — it locks itself with respect to s. However, any signal different from s is again retransmitted and afterwards the node locks itself again with respect to that last signal, etc. Thus, a locked node remains in the receiving mode until it gets unlocked by a different signal.

Again, in [7] is is shown that given any $\varepsilon : 0 < \varepsilon < 1$, Algorithm Broadcast delivers s to each node of $\mathcal{N}$ that has not been locked with respect to s in time $O(DQ\log(n/\varepsilon))$ and with probability $1 - \varepsilon$. Afterwards, all nodes in $\mathcal{N}$ will be in a locked state with respect to s.

In order to achieve the failure probability ε of the broadcasting algorithm *Protocol Send* must be performed with error probability ε/n. This calls for repeating each send operation in a node $k = O(Q\log(n/\varepsilon))$ times—i.e., this time k should grow not only with a decreasing ε, but also with increasing size of the nanonet. Similarly as in the case of *Protocol Send,* the respective timer is realized by means of the timing mechanism of the underlying automata.

Now it is time to return to the problem of non well-behaved nanonets. From the description of the communication protocol it is seen that occasional short-time connectivity "interruptions" could not harm the communication as long as constant k controlling the repeated sending trials is set high enough to overcome the interruption periods. A sufficiently high value of k will also help to overcome differences and drifts in the clock rates of the individual nodes, the switching times between sending and receiving modes, and occasional local increase of maximal neighborhood size which can block efficient communication. Finally, a generous estimate of the nanonet diameter (which is always bounded by n)

that is in fact determined by the shape of the closed environment in which the nanomachines operate will help to overcome the time variations of instantaneous nanonet diameter.

4   Distributed Computing through Nanomachines

Let us assume that each nanomachine gets its "own" input from the domain $\{0, 1\}$ through its sensors; this input can be used in the subsequent "collective" data processing. If outputs from the nanomachines are also restricted to the domain $\{0, 1\}$ then the respective nanonet can be seen as a device computing functions with their inputs and outputs represented in unary. Further assume that within the net there is a distinguished nanomachine called the *base*. In the full version of the paper [8] it is shown that under this scenario a nanonet can simulate a very simple model of a universal (i.e., programmable) machine called *counter machine* (cf. [6]), with high probability. A counter machine computes with the help of a finite number of so-called counters. These counters represent any positive integer number and the only allowable operation over counters is their testing for zero, and adding or subtracting a one (the later operations can only be used for counters representing a positive number).

In a simulating nanonet each counter storing a number $n \geq 0$ is represented by a set of nanomachines of cardinality exactly n whose states belong to a specific subset of states. To simplify matters, assume that all machines in a counter are in the same state, q, let us say. Testing such a counter for zero is easy: using *Algorithm Broadcast* the base station issues a "query" whether there is a nanomachine in state q. Using the same algorithm each machine in state q sends the answer "yes" which eventually, in time that can be computed from the estimates given in Section 3, will reach the base station. If no answer arrives in the given time, than the counter is zero.

In order to add a one to a counter represented by a set S of nanomachines in state q we must select a single machine that is not in state q to be added to S by changing the state of that nanomachine to q. To subtract a one from a counter represented by S we must select a single machine in S and change its state to a state different from q. In both cases, a single machine has to be selected from a set of machines. This operation is called leader selection. The *leader* of a set of nanonmachines is a single nanomachine which originally has belonged to that set but subsequently is put into a distinguished state that is different from the states of all other nanomachines in that set.

The algorithm for leader selection from among the candidate set of nanomachines works in rounds. The idea of the algorithm is as follows. Using their probabilistic mechanism, in each round the nanomachines from the candidate set throw a random coin giving output 0 or 1. If this will split the set of candidate machines into two non-empty subsets we proceed recursively with either subset. Otherwise we choose the non-empty set which with a high probability is a singleton set—the leader. The instructions "what to do" are sent by the base station using *Algorithm Broadcast*. For the details and complexity analysis, see the full paper [8]. The resulting simulation algorithm is unbelievably cumbersome since all the computation is performed in a unary counting system and the operations have to be repeated many times in order to reach a prescribed level of reliability. In [8] the following theorem is shown:

Theorem 1. *Let $\mathcal{M}$ be a counter machine with input of size n operating with a constant number of registers of size $O(n)$ in time $p(n)$. Then for any $\varepsilon : 0 < \varepsilon < 1$ there exist constants $c_1 > 0$ and $c_2 > 0$ and a nanonet $\mathcal{N}$ consisting of $O(n)$ nanomachines using* Protocol Send *with failure probability ε such that $\mathcal{N}$ simulates $\mathcal{M}$ for inputs of size at most n in expected time $O(DQp(n)\log n\log(n/\varepsilon))$ with probability of failure bounded by $\max\{1, c_2 p(n)(\varepsilon \log n + (\varepsilon/n)^{c_1})\}$.*

Thus, the theorem claims that a nanonet of a fixed size can correctly, with a high probability, simulate any computation over inputs up to a certain size. For larger inputs a larger nanonet must be used. Most probably, in practice nobody would consider performing a universal computation by nanonets. Nevertheless, our result shows that in principle a nanonet can perform whatever algorithmic (or nano-robotic, if the machines embodiment is considered) task arising in practical applications.

References

1. Adar, R., Benenson, Y., Linshiz, G., Rozner, A., Tishby, N., Shapiro, E.: Stochastic computing with biomolecular automata. Proc. Natl. Acad. Sci. USA 101, 9960–9965 (2004)
2. Alur, R., Dill, D.L.: A Theory of Timed Automata. Theoretical Computer Science 126, 183–235 (1994)
3. Bath, J., Turberfield, A.J.: DNA nanomachines. Nature nanotechnology 2, 275–284 (2007)
4. Hiyama, S., Moritani, Y., Suda, T., Egashira, R., Enomoto, A., Moore, M., Nakano, T.: Molecular Communication. In: Proc. of the 2005 NSTI Nanotechnology Conference (2005)
5. Kwiatkowska, M., Norman, G., Parker, D., Sproston, J.: Performance Analysis of Probabilistic Timed Automata using Digital Clocks. In: Formal Methods in System Design, vol. 29, pp. 37–78. Springer, Heidelberg (2006)
6. Minsky, M.: Computation: Finite and Infinite Machines. Prentice-Hall, Englewood Cliffs (1967)
7. Wiedermann, J., Petrů, L.: On the Universal Computing Power of Amorphous Computing Systems. Published online in Theory of Computing Systems, January 27. Springer, New York (2009)
8. Wiedermann, J., Petrů, L.: Communicating Mobile Nano-Machines and Their Computational Power. Technical Report V-1024, Institute of Computer Science, Prague, May 2008, accessible from the web pages of the institute

Author Index